Curing Concrete

Curing Concrete

PETER C. TAYLOR

CRC Press
Taylor & Francis Group
Boca Raton London New York

CRC Press is an imprint of the
Taylor & Francis Group, an **informa** business

CRC Press
Taylor & Francis Group
6000 Broken Sound Parkway NW, Suite 300
Boca Raton, FL 33487-2742

First issued in paperback 2019

© 2014 by Peter C. Taylor
CRC Press is an imprint of Taylor & Francis Group, an Informa business

No claim to original U.S. Government works

ISBN-13: 978-0-415-77952-4 (hbk)
ISBN-13: 978-0-367-86556-6 (pbk)

Library of Congress Cataloging-in-Publication Data

Taylor, Peter C. (Peter Clement)
  Curing concrete / Peter C. Taylor.
    pages cm
  "A CRC title."
  Includes bibliographical references and index.
  ISBN 978-0-415-77952-4 (hardcover : alk. paper)
    1. Concrete--Curing--Handbooks, manuals, etc. 2. Concrete construction--Handbooks, manuals, etc. I. Title.

  TA440.T285 2014
  620.1'36--dc23   2013026193

Visit the Taylor & Francis Web site at
http://www.taylorandfrancis.com

and the CRC Press Web site at
http://www.crcpress.com

To my family
This is why we took as many photos of concrete
as mountains on our road trips.

Contents

List of figures

List of tables

Preface

Students are taught that curing concrete is important, but in practice curing is often a low priority on the construction site. This is most likely because the benefits of the cost of curing are not immediately apparent, and the consequences of poor curing may only appear later in the life of the structure.

The fundamental principle behind the curing of concrete is simple: The mixture should be kept warm and wet for several days after placement in order to achieve the properties needed. This is because cement hydration is a relatively slow process that requires sufficient water available to continue. Drying normally occurs at the surface, meaning that poor curing affects the surface by reducing resistance to the environment and abrasion; yet this is precisely the zone that is exposed to bad weather and tires.

Demands on modern concrete are increasing, while raw materials are changing and budgets are shrinking, together requiring that closer attention is needed to ensure that the best value is obtained from the cementitious materials in a mixture.

The aim of this book is to help those involved with working in concrete construction understand why curing is important, that it is indeed possible and worth the effort, and to show how it can best be carried out.

The discussion includes the fundamentals behind hydration and why curing is needed; how properties are affected; and how curing can be effectively specified, provided, and measured. The final chapter includes a review of published work evaluating curing in real-world structures.

Acknowledgments

I first have to acknowledge my mentors and colleagues who, through the years, have taught me about engineering, concrete, and balancing lab theory and site practice. These include people from the University of Cape Town, Cement and Concrete Institute South Africa, CTLGroup, PCA, CP Tech Center, and the U.S. paving construction industry. I am especially indebted to Brian Addis and Rachel Detwiler for teaching me how to write concisely and precisely. Thank them for this volume being short, sweet, and to the point.

A significant amount of material came from two sources: reviews of curing practice by Kenneth W. Meeks ("Curing of High-Performance Concrete: Report of the State-of-the-Art," National Institute of Standards and Technology) and Toy S. Poole ("Curing Portland Cement Concrete Pavements, Vol. II"). Other references were obtained through the Iowa State University Library. (Life is so much easier when you can search a collection electronically from the safety and comfort of your own home. How did we do it before?)

Jim Grove and J. Z. Zhang were very responsive to my phone calls begging for photographs. More than once they jumped in their cars to go find that perfect image for me. Professor Mark Alexander spent several hours recovering an old report he had just thrown away, just so I could quote it correctly.

All of my colleagues have put up with me bouncing ideas and droning on, and in so doing helped to ensure that the following pages are worth reading and learning from.

Finally, my wife has endured many evenings of being ignored while I stared at a computer screen. Need I say more?

About the author

Peter C. Taylor was born and educated in Southern Africa, earning a BSc and PhD at the University of Cape Town. He spent 10 years with consultants in Chicago, involved in a wide range of consulting and research projects that investigated the effects of raw materials on concrete performance, both in the laboratory and in the field.

Dr. Taylor has been employed at the National Concrete Pavement Technology Center at Iowa State University since 2007. During that time he has been involved in managing and conducting research projects and programs investigating and teaching materials related to aspects of concrete pavements. His primary area of interest is in concrete durability, including the development of test methods for assessing the potential durability of concrete mixtures and the implementation of approaches to improve longevity. Dr. Taylor's responsibilities include developing technical resources, such as manuals on materials aspects of concrete mixtures. He also travels throughout the United States teaching practitioners about the latest technologies available to improve pavement reliability and sustainability.

Dr. Taylor is the author of more than 50 publications and is an active member of Transportation Research Board (TRB), ASTM International, and American Concrete Institute (ACI) committees.

Chapter 1

Introduction

Curing is one of those activities that every civil engineer and construction worker has heard about, but in reality does not worry about much. Many publications lead with statements about how curing is critically important, yet in practice curing is often low on the list of priorities on the construction site.

The aim of this publication is to help those involved in working in concrete construction understand why curing is important, that it is indeed possible and worth the effort, and to show how it can best be carried out.

The basic need for curing of concrete is very simple. Cement hydration is a chemical process that requires the presence of water over a relatively long period of time at a reasonable range of temperatures. Curing is defined as "action taken to maintain moisture and temperature conditions in a freshly placed cementitious mixture to allow hydraulic cement hydration and (if applicable) pozzolanic reactions to occur so that the potential properties of the mixture may develop" (American Concrete Institute 2010). In other words, curing is the work we do to encourage concrete hydration until we get the properties that we want in the mixture.

A key factor discussed is that moisture curing only affects the outer 30 to 50 mm of the surface of a concrete element. This means that moisture control is not primarily for the purpose of enhancing compressive strength of a structure. On the other hand, it is enormously influential on surface permeability and hardness, so it effectively controls the potential longevity of a system, particularly those exposed to severe environments.

Tied to this is the increasingly common use of mixtures with water–cement ratios below 0.4, which are at greater risk of self-desiccation. Although providing external water is critical to obtain the full benefits of such a mixture at the outer surface, it will not help more than about 30 mm below the surface. In this case consideration may be given to using internal curing techniques.

Another growing trend is the use of supplementary cementitious materials (SCMs), most of which tend to hydrate more slowly and for longer. A well-hydrated mixture containing SCMs will therefore generally exhibit far better potential durability in the long term, but a mixture containing SCMs that has been allowed to dry prematurely will likely be worse off than a similarly treated plain cement mixture.

Another key factor is that curing includes work to control concrete temperature. Structures or slabs that are too cold in the first few hours after placement will hydrate very slowly, if at all. This may lead to the need to leave forms in place for longer or provide protective systems to prevent plastic shrinkage cracking. On the other hand, concrete that is permitted to get too hot in the first few hours is likely to be at greater risk of cracking due to differentials between internal and surface temperatures. Concrete that hydrates at higher temperatures is also more prone to deleterious chemical reactions and attack by external chemical reagents.

Although it may be accepted that curing is a necessary task, it also imposes a cost and may cause delays to the construction program. It is also perceived that a lack of curing does not visibly benefit the concrete quality or affect pay factors tied to the concrete. It is therefore easy to ignore the need to cure when budgets and timelines are under pressure. The increasing demands being placed on concrete mixtures mean that they are less forgiving than in the past. Therefore, any activity that will help improve hydration and so performance, while reducing risk of cracking, is becoming more important than ever.

Finally, there is a need to measure whether curing has been carried out properly. Direct approaches such as measuring the volume of curing compound per unit area are simple to monitor but do not take into account local variations, wind losses, or whether it was carried out in time. If techniques other than curing compounds are utilized, then the question arises about how long they are required. Performance-based approaches to answer these questions have some appeal but tend to be expensive and imprecise.

Tied to the issue of measurement is that of payment. Some contracts have a line item that makes curing a pay item, but without effective measurement it is difficult to assess what payment is due, and often the cash value attached to this is far less than the potential harm that can be inflicted on the structure if it is ignored or conducted poorly. In theory, curing is a good thing, but the reality is that we need to balance the required performance of the system with the effort and cost it takes to get there.

The following chapters will help the reader to understand why curing is important, how it can best be carried out, and what specifications and approaches are available to improve the probability that it is done properly.

Chapter 2 discusses the hydration of cementitious systems and lays the groundwork for understanding why curing is a critical activity when working with concrete. Chapter 3 describes how curing activities will influence and change the properties of concrete that are critical to its long-term performance in a variety of applications. Chapter 4 describes different technologies and techniques available to encourage hydration and discusses their applicability in different applications. Chapter 5 summarizes specifications that describe requirements for materials used in curing as well as outlines what is needed in specifications that will address all the details of how curing can be called out in a contract. Most of the specifications discussed are U.S. based and refer to ASTM or American Concrete Institute documents, but reference has been made to European Standard (or EN) documents where appropriate. The final chapter provides worked examples from published literature of how curing, or the lack thereof, has affected concrete performance in real-world applications.

There is a great need to educate all people involved in a project, from the owners' board of directors to the person stripping forms, that curing is actually good value for money. Allowing cement to be unhydrated for want of a bit of water, so compromising the quality and longevity of a large infrastructure system, does not seem logical. It is hoped that this volume will help in driving that education process forward.

REFERENCE

American Concrete Institute (ACI), 2010, "ACI Concrete Terminology," American Concrete Institute, Farmington Hills, MI, http://terminology.concrete.org (accessed September 2012).

Chapter 2

Cementitious materials: Chemistry and hydration

INTRODUCTION

At the heart of every concrete mixture is portland cement, a seemingly boring gray powder that is in fact extremely complex. It is used in every country in the world and is the backbone of the infrastructure that makes our present civilization and lifestyle possible.

Fundamental to this success is that cement and cementitious materials go through a chemical process when added to water that turns a pile of rock, sand, and powders into strong, durable structures that carry enormous loads and traffic. What is remarkable is that much of the concrete used is made, formed, and finished on an exposed construction site rather than in a tightly controlled factory setting. This is despite the fact that the long-term performance of any structure can be strongly affected by the environment that it is exposed to in the first few hours after placing.

The real need is to do everything we can to ensure that the best value and performance is achieved from the mixture, and effective curing can make a big difference in this endeavor.

This chapter provides a brief background on why concrete works, and what has to be done to make it work well.

Concrete comprises four basic sets of ingredients:

- Cementitious materials, including portland cement and supplementary cementitious materials such as fly ash and slag cement
- Water
- Chemical admixtures
- Aggregates

The reactions between the first two sets, known as hydration, provide the primary compounds that bind concrete together. Hydration is an initially rapid but gradually slowing chemical reaction that requires water to be present for it to continue, thus explaining the fundamental need for curing. The following sections describe each of the materials used in concrete mixtures and how they affect mixture performance over time.

PORTLAND CEMENT

Cement is a generic word meaning "glue"; portland cement is a specific type of glue that is used in the construction of civil infrastructure. The chemical composition of portland cement typically falls within a fairly tight window, as illustrated in Table 2.1 (Kosmatka and Wilson 2011).

The interesting thing about portland cement is that it is hydraulic, meaning that it hardens and gains strength in the presence of water. This is counterintuitive when many hardening materials have to be allowed to dry out to be effective. This property is particularly useful because properties such as strength and impermeability will continue to improve in a system that is kept wet over a long period of time, providing some additional factor of safety in structures and a limited ability to self-heal small cracks, particularly in water-retaining structures.

Portland cement is mostly comprised of four compounds that are combinations of calcium oxide, silica, alumina, and iron oxide. The combinations of these oxides are complex and nonuniform providing ample opportunity for research. From a practical point of view, the four compounds may be considered as follows (Figure 2.1; Taylor et al. 2006b). Note that cement chemists use a unique notation shown in Table 2.2.

Silicates
    Alite, C_3S
    Belite, C_2S
Aluminates
    Tricalcium aluminate, C_3A
    Calcium alumino ferrite, C_4AF

Calcium sulfate compounds ($\bar{S}$) are also added to the system in the form of gypsum in order to control hydration rates of the aluminate phases.

Table 2.1 Range of Portland Cement Chemical Composition

Oxide	Percent by Mass
CaO	60.6–66.3
SiO_2	18.6–23.4
Al_2O_3	2.4–6.3
Fe_2O_3	1.3–6.1
SO_3	1.8–4.6
MgO	0.6–4.8
Na_2O_{eq}	0.05–1.20

Source: Kosmatka, S. H., and Wilson, M. L., *Design and Control of Concrete Mixtures*, 15th ed., EB001, Portland Cement Association, Skokie, IL, 2011. With permission.

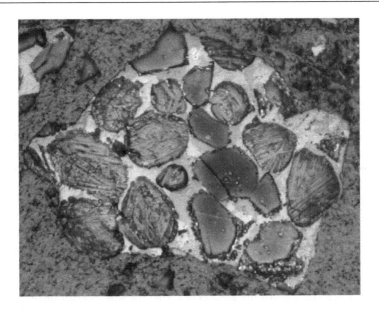

Figure 2.1 A micrograph of a cement grain; etched polished section viewed in reflected light. (From CTLGroup. With permission.)

Table 2.2 Cement Chemistry Notation

Oxide	Standard Notation	Cement Notation
Calcium oxide	CaO	C
Silica	SiO_2	S
Alumina	Al_2O_3	A
Ferrite	Fe_2O_3	F
Sulfate	SO_3	$\bar{S}$
Water	H_2O	H

Source: Kosmatka, S. H., and Wilson, M. L., *Design and Control of Concrete Mixtures*, 15th ed., EB001, Portland Cement Association, Skokie, IL, 2011. With permission.

An x-ray fluorescence (XRF) analysis of a sample of cement will report the elements in the sample, normally expressed as oxides. It is unlikely that any of the elements present are actually present in the form of oxides, but this approach does allow analysts to evaluate cements based on these numbers. A set of equations is available to calculate, from the oxide analysis, the so-called Bogue compounds that express the chemistry of the cement as percentages of the aforementioned minerals. Again, this approach is theoretical and unlikely to exactly portray the system, but does help in predicting the performance of the cement. The alternative is to conduct quantitative x-ray diffraction analyses to determine the amounts of the different compounds.

The hydration products of the silicates are different forms of calcium, silica, and water, generically known as calcium-silicate hydrate (C-S-H). The different forms vary by the relative amounts of calcium and silica in them. These are the compounds that provide the mixture with the strength and impermeability that makes concrete useful in construction. C-S-H looks like thin fingers that start to grow out of the cement grain when hydration starts. When the fingers interlock, the mixture starts to stiffen and changes from liquid to solid. As hydration proceeds the spaces fill in with more C-S-H to form a solid mass. An additional hydration product is in the form of hexagonal calcium hydroxide crystals (Figure 2.2; Kosmatka and Wilson 2011).

The hydration product of C_3A and sulfate is initially ettringite that within a few hours converts to monosulfate with continued reaction. Ettringite formation can lead to early stiffening and contribute to some of the initial strength.

C_4AF does not contribute significantly to the engineering properties of the mixture, but increasing amounts of C_4AF will lead to a cement that is darker in color.

More information about the hydration process is provided in a later section.

Figure 2.2 A micrograph of a partially hydrated cement grain showing C-S-H (fingers and flowers) and CH (hexagonal plates). (From Kosmatka, S. H., and Wilson, M. L., *Design and Control of Concrete Mixtures*, 15th ed., EB001, Portland Cement Association, Skokie, IL, 2011. With permission.)

SUPPLEMENTARY CEMENTITIOUS MATERIALS

Supplementary cementitious materials (SCMs) are a broad class of materials that are normally the by-product of other industrial processes. They react with portland cement to enhance the properties of concrete containing them.

Their primary benefit is that silica in them reacts with the calcium hydroxide that is a result of the cement hydration, to form calcium silicate hydrate. This in effect converts a compound that contributes little to the structural integrity of the system into a primary part of the binding system. This is known as the pozzolanic reaction. If there is sufficient calcium in the material and the alkalinity of the mixture is high enough, then some supplementary cementitious materials will set and harden on their own, and are therefore considered cementitious.

Predominant in terms of usage are fly ash, slag cement, and silica fume.

Fly ash is a by-product of burning finely ground coal for generating electricity. Such facilities commonly use relatively low-grade coal, meaning that it contains significant amounts of rock. This material melts in the furnace and is blown up the stack, rapidly cooling to form small glassy spheres that are basically mixtures of calcium, silicon, iron, and aluminum compounds. The particles are captured in bag filters or electrostatic precipitators before the gases are emitted to the atmosphere. Fly ash particles are typically about the same size as cement particles; in the range 1 to 100 μm. The size and glassy nature of these particles makes them soluble enough that they will react, albeit relatively slowly, in a cementitious system. Bottom ash that is allowed to drop to the bottom of the furnace cools slowly and is therefore not glassy, and is not useful as a pozzolan.

The composition of fly ash will vary from plant to plant due to differences in the source coal and its bedrock, and in processing practices. The bulk of most fly ashes comprise minerals containing oxides of iron, silica, alumina, and calcium. Smaller amounts of magnesium, sulfur, sodium, potassium, and carbon are normally detected. Generally, the lower the calcium content the slower the reactivity (leading to slower strength gain), but the better the long-term performance in terms of potential durability, if the system is kept wet and warm. Some unburned carbon may remain in the system, often in the form of randomly shaped and porous particles. These particles preferentially absorb air entraining admixtures, making it difficult to control air contents in fresh concrete. The amount of this material is therefore limited by U.S. code to 6%, and in practice to about 3% or 4%, expressed as a loss-on-ignition when heated to 950°C.

Workability of mixtures containing fly ash is generally improved although some increase in stickiness may be observed. Setting times are normally extended because reaction rates are slower than those of portland cement. For the same reason the initial strength gain is normally slower, although it will continue for longer leading to better performance in the long term. Some fly ashes act in a cementitious manner as well as pozzolanic, meaning that higher dosages can be used without overly affecting setting time and strength gain. Mixtures containing fly ash will exhibit significantly lower permeability, particularly in the long term, thus leading to improved potential durability. Some fly ashes are also effective in reducing the risk of alkali silica reaction and sulfate attack, depending on system chemistry, dosage, and degree of hydration. Fly ash has been the subject of extensive research since the 1940s and there is a large body of literature discussing its properties and usage (American Concrete Institute [ACI] Committee 232 2003; Thomas 2013).

Slag cement is the molten material that is drained off a blast furnace used to process iron ore into iron. The molten rock is dropped through a cool air stream or water stream, again causing it to freeze rapidly in a glassy form. The material is then ground to a fineness similar to cement. Slag cement generally has a relatively high calcium content, meaning that it will behave somewhat like a hydraulic cement, particularly in a highly alkaline environment. Slag is harder than cement; therefore if it is interground in cement it will normally have coarser particles than the cement. When used at low dosages, mixtures containing slag perform in much the same way as plain cement mixtures. At higher dosages, setting times and strength gain may, again, be slower but continue for longer (ACI Committee 233 2003).

Silica fume is a by-product of the ferro-silicon industry and is the very fine particulate material emitted from hearth-type kilns. It comprises mostly silica, also in the form of glassy spheres, and is generally about 100 times finer than portland cement. To facilitate handling it is commonly tumbled in an air stream causing the particles to agglomerate into bundles about the size of a cement grain held together by electrostatic forces. The extreme fineness makes the material very reactive, leading to rapid strength development and very low permeability in concrete. On the other hand, the extreme fineness tends to stop bleeding, increasing the risk of plastic shrinkage cracking in flat slabs. Such elements containing silica fume must be well protected from drying as soon as it is placed (ACI Committee 234 2003).

The effects of duration of moist curing were studied on three concrete mixtures containing 2.5%, 5%, and 10% silica fume (Marusin 1989) with water–cement ratios (w/cm) from 0.37 to 0.34. Moist curing was provided by storing 4-inch samples in plastic bags with wet sponges for selected ages. The cubes were then air dried for 21 days before soaking in salt solution for 21 days, followed by further air drying for another 21 days. Chloride ion penetration was determined at four depths. Chloride ion penetration decreased with increasing curing length up to the maximum 21 days measured. Other work has shown that after about 56 days, silica-fume concrete will gain little additional strength, probably due to the effects of self-desiccation (Hooton 1993).

More information about these materials and other pozzolans such as metakaolin and rice husk ash are available in the literature (American Coal Ash Association [ACAA] 2003; ACI Committee 232 2003; ACI Committee 233 2003; Holland 2005; Siddique and Khan 2011).

All of the materials make a positive contribution to the sustainability of a structure because they are by-products from other industries, so reducing the need for landfill, and they replace some of the portland cement in the mixture, so reducing the energy burden and CO_2 footprint associated with a given mixture.

In broad terms, the materials most commonly used as supplementary cementitious materials will slow the rate of hydration for the first few days, making the mixture more sensitive to early drying. Moisture control must therefore be more rigorous and potentially be applied for a few days longer if mixtures contain supplementary cementitious materials. The balance to this is that performance of mixtures containing supplementary cementitious will tend to continue for longer leading to better performance in the long term.

HYDRATION

The chemical reaction between cement and water is known as hydration. It begins as soon as water and cement are mixed, and proceeds through several stages over time as illustrated in Figure 2.3 (Taylor et al. 2006b). The

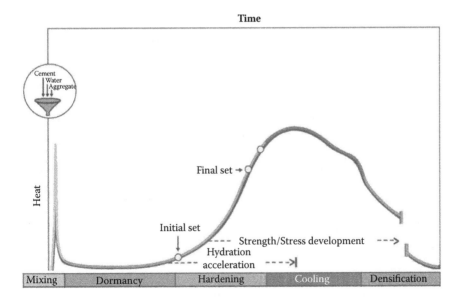

Figure 2.3 Hydration chart. (From CPTech. With permission.)

reactions of these compounds with water can be described in five stages: mixing, dormancy, hardening, cooling, and densification.

During mixing, the primary action is the start of the dissolution of the cement grains releasing various ions into the water. The aluminates and gypsum dissolve within minutes of being exposed to the water and react to form calcium-alumina-sulfate hydrates very rapidly. This reaction is strongly exothermic and energy is released fast, but because it occurs over a relatively short period, a temperature increase is normally only observed in small paste samples. The calcium-alumina-sulfate hydrates create a gel-like substance that coats the cement particles so limiting the ability of water to reach the cement to continue dissolving it. This effectively slows dissolution and hydration to negligable rates, setting up the second stage known as dormancy.

The reactions are broadly represented as (Kosmatka and Wilson 2011):

$$2C_3A + 3C\bar{S}H_2 + 26H \rightarrow C_6A\bar{S}_3H_{32} \text{ (ettringite)}$$

$$2C_3A + C_6A\bar{S}_3H_{32} + 4H \rightarrow 3C_4A\bar{S}_3H_{12} \text{ (monosulfate)}$$

If C_3A is exposed to water without sufficient sulfates in solution, the reaction to form a calcium-alumino-hydrate is extremely rapid and exothermic, leading to what is known as flash set. On the other hand, excess sulfates in solution will lead to temporary false set. Cement manufacturers go to considerable lengths to ensure that the right amount of sulfate is in their cement. If flash (permanent and premature) set occurs, expert advice should be sought (Taylor et al. 2006a).

It is believed that the early reaction products of C_3A, $\bar{S}$, and water coat the cement particles thus limiting the rate of reaction and controlling setting times. This period over which the reactions are suppressed for about two to four hours is useful because it allows time for the concrete to be thoroughly mixed, transported, placed in its final position, and finished to the desired surface. During this time, the concrete is plastic and does not generate heat. Although it appears that nothing is happening in the mixture, the cement is continuing to slowly dissolve, and the water is becoming saturated with dissolved calcium and OH⁻ (hydroxyl) ions.

Stiffening and hardening begins when the water becomes supersaturated with dissolved calcium ions. When this occurs, it triggers formation of calcium-silicate-hydrate (C-S-H) and calcium hydroxide (CH) compounds in another set of exothermic silicate reactions. It is also believed that at the end of the dormant period, the shells around the cement grains are burst open, rapidly releasing additional dissolved ions into the system resulting in a sudden acceleration in reaction rates and a marked increase in temperature. The C-S-H products are finger- or petal-like growths (Figure 2.4; Sutter 2008) while the CH is crystalline (Figure 2.2). These compounds interweave and mesh together around the aggregates, causing the concrete

Figure 2.4 C-S-H imaged in an environmental SEM showing a petal-like morphology. (Courtesy of Lawrence Sutter. With permission.)

to become stiffer and eventually a solid. Initial set is when the interference of the hydration products causes the mixture to begin stiffening and is exhibited as an increase in yield stress or in the force required to press an object into the surface, or when the speed of sound through the mixture begins to increase. It is also loosely associated with the time when the temperature of the mixture begins to rise.

The silicate hydration reactions can be represented as (Kosmatka and Wilson 2011):

$$2C_2S + 9H \rightarrow C_3S_2H_8 + CH$$

$$2C_3S + 11H \rightarrow C_3S_2H_8 + 3CH$$

It is notable that the C-S-H (shown in the equations as $C_3S_2H_8$) is the same form from both sets of reactions. The differences are in the amount of CH with considerably more being formed from C_3S than C_2S. The CH does not contribute significantly to the mechanical properties of the system but is available to react with supplementary cementitious materials as discussed later.

C_2S hydration also tends to be slower than C_3S meaning that water has to be available for a long period for these compounds to react if the full benefit of the cement is to be realized.

It should be noted that temperature increase is a measure of when the silicate reactions accelerate, and there may be a variable interval between the beginning of these reactions and when sufficient hydration product is formed to cause physical interactions between particles. Final set is somewhat arbitrarily defined, therefore correlating standard test procedures based on pressure with other parameters such as temperature or speed of sound are difficult.

During this time the aluminates (C_3A) continue to react with gypsum and water to form ettringite, a thin needle-like formation that will also contribute to early stiffening and some strength gain.

After several hours, sufficient hydration products have formed to once again limit the access of water to the undissolved cement. Hydration therefore begins to slow as indicated by the drop in temperature.

A mixture will functionally change from liquid to solid while it is near its hottest and therefore at its greatest volume. Subsequent cooling as heat of hydration drops, often at the same time as ambient cooling, can set up significant stresses. These stresses occur when the mixture is gaining stiffness rapidly, but has yet to have significant strength, thus making this the time when the risk of cracking is at a maximum. This can be accommodated by joint sawing in the case of slabs on grade. In the case of mass concrete it is the differential temperature between the interior and the surface that will set up stresses that can be addressed by cooling systems using embedded pipes to cool the interior or by using blankets to prevent rapid cooling at the surface.

Over time the hydration continues, as long as there is sufficient cement and water, with reaction products filling the spaces between the fingers eventually to form a solid mass that is strong and impermeable (Figure 2.5). The initial setting, hardening, and early strength are mainly as a result of the reactions of the C_3S portion of the cement. The C_2S reactions are much slower and tend to contribute more to concrete performance after several days.

Supplementary cementitious materials will also start to dissolve when mixed with the water. The silicates in them will react with calcium hydroxide to produce more C-S-H, which adds to system density, impermeability, and strength. Some aluminates if present may also react to form hydration products that may add to stiffening and early strength. SCM reactions are accelerated in the presence of alkalis, and when the system is heated such as in steam curing. The final performance of a well-proportioned, well-cured SCM mixture will generally be significantly better than a similar plain cement mixture.

As noted before, supplementary cementitious materials will tend to react more slowly but will continue for longer, leading to greater strength and impermeability over time, as long as there is sufficient water for the reactions to continue (Figure 2.6; Kosmatka and Wilson 2011). It is a reasonable rule of thumb that systems that are permitted or forced to react relatively slowly initially will tend to continue for a longer time and in the long term

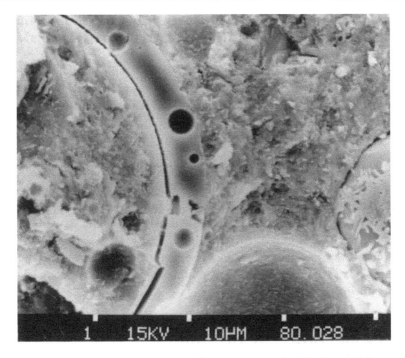

Figure 2.5 Dense hydrated cement paste after several months of hydration. Note some large unhydrated fly ash particles in the system.

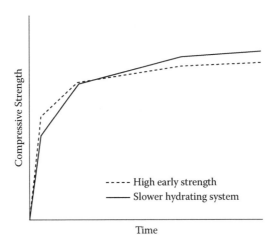

Figure 2.6 Systems that hydrate rapidly early will tend to slow sooner, leading to reduced long-term performance.

provide better quality. Conversely, systems that are accelerated early on will tend to reach a maximum and flatten out with little added improvement after about a month.

Concretes containing silica fume, on the other hand, tend to hydrate rapidly initially, leading to high strengths and very low permeability. This is due to the combination of the relatively high solubility of the very small particles as well as their effect on nucleation of hydration products. Silica-fume concrete is reported to be more sensitive to a cold environment than concrete without silica fume (Cabrera et al., 1995). This is likely because the lower temperatures slowed hydration more markedly in the silica fume systems.

A significant point to note in the hydration chart is that the rate of heat generation, indicating rate of reaction, does not return to zero for a long period of time in a system that has sufficient reagents and space for reaction products to grow into. Although there is a lot of activity in the first 24 hours, even the first week, the significant development of hydration products leading to impermeability, and therefore high potential durability, occurs during the first months.

MIX DESIGN AND PROPORTIONING

Mix design is a broad term that has a multitude of meanings. A definition that is being adopted is that *mix design* is the process of choosing what is required of a mixture in terms of fresh and hardened properties, whereas *mix proportioning* is the process of choosing the materials and their proportions to achieve the requirements of the design (Taylor et al. 2006b).

The current contractual processes are prone to setting up disputes, because in some cases, decisions made in the mix design stage are made by the wrong person. An example would be specifications that place limits on workability, when it may not be clear at this stage whether the concrete will be placed by pump or by bucket. Implicit behind this process is that the contractor and the owner are primarily concerned about very different things. The contractor is strongly influenced by the fresh properties of the mixture: the ease with which it can be batched, transported, placed, and finished. On the other hand, the owner is concerned about how long the concrete will last under the loads and environment to which it is exposed. These requirements may have mutually exclusive solutions, requiring that a compromise has to be made on both sides. Examples of who should be making some selections include the following:

- Strength, air-void system parameters, and w/cm are generally considered to be the engineer's choice in order to help ensure performance measures.

- Slump is often specified by the engineer, although it has no direct effect on concrete quality and should be based on the placement method and equipment, which may not be known at the design stage.
- SCM percentage may be based on performance requirements such as alkali silica reaction, or on economic and constructability requirements.
- Cement content does not directly govern performance for a given w/cm (Yurdakul 2010). It should therefore be determined at the proportioning stage to achieve the required performance using the cementitious system selected, most economically within the constraints provided by the aggregate system.
- Aggregate grading. Although a good combined grading is preferred normally to improve workability, it is still possible to make good concrete with a marginal grading.
- Admixture dosage should be based on mixture properties and environment and not on a fixed number.

Once the mix design parameters have been settled upon, the proportioning exercise may begin. There are a number of published approaches to this activity. The ACI Committee 211 report (1991) uses a volumetric approach that begins with a factor based on the fineness modulus of the fine aggregate and requires several iterations. The approach used by Shilstone (1990) starts with the combined aggregate grading and uses charts to assess whether the selections are acceptable. Although there is logic behind these approaches, they are fundamentally empirical. Another approach suggested by ICAR for Self Consolidating Concrete (Koehler and Fowler 2010) is to

- Choose aggregate system
- Choose paste quantity
- Choose paste quality

Using this methodology, the following steps may be considered (Grove and Taylor 2012):

- Combine aggregates to maximize their volume; recognizing that some extra paste is required to lubricate the mixture and achieve workability. This can be assessed by using the tools described by Shilstone (1990; see also Taylor et al. 2006b). Effort is still needed to rationalize this approach because maximum density my not be ideal for the workability required, particularly in fluid systems such as self-consolidating concrete.
- Choose the cementitious system based on a balance of specification limits, durability, and other performance requirements, cost, and availability.
- Choose w/cm, again to achieve the performance requirements using the cementitious system selected.

- Choose air content based on the likelihood of freezing cycles and saturation.
- Add sufficient paste to fill all the voids and coat the aggregates, and a bit more to achieve workability. Details of determining this amount are under research.
- Add admixtures to control final workability and setting characteristics.

This will lead to selection of a series of numbers. They are only useful to guide preparation of trial mixtures to gauge whether the mixture actually performs as desired. This is because there are still some factors that will influence both fresh and hardened properties that are inadequately modeled with current knowhow. Such factors include the influence of cementitious chemistry on strength gain and durability; the influence of aggregate shape and texture on workability and strength; and the definition and measurement of "workability."

Trial batches should be conducted initially in small batches to gauge overall performance, then in full scale and preferably field trials to take into account effects of batch plant size and mixing efficiency, and the effects of the weather that are not observed in a temperature controlled laboratory.

Materials should not be substituted without additional trial mixes, because the specifications for most materials are broad enough that materials of the same classification may exhibit significantly different performance, potentially leading to severe problems in the fresh and the hardened concrete.

The benefits and effects of the different ingredients in a mixture are summarized in the following sections. More detailed information is available in the literature (Kosmatka and Wilson 2011; Taylor et al. 2006b). Understanding these interactions helps the designer to select the right materials and appropriate dosages.

Cements

The effects of changes in cement chemistry on concrete mixtures are summarized in Table 2.3 (Johansen and Taylor 2005). Slower hydrating systems will generally have to be protected from the environment for a longer period to prevent moisture loss and consequent early cracking and to ensure adequate development of engineering properties such as strength and impermeability.

Supplementary cementitious materials

In general, the effects of using supplementary cementitious materials include

- A reduction of the amount of water required to achieve a given workability. This is particularly marked with fly ash depending on the

Table 2.3 Effects of Changes in Cement Characteristics on Concrete Properties

	Influence on Concrete (Assuming Only a Change in the Given Component and Nothing Else)
Alkali (increasing) Clinker sulfate (increasing)	Air content increases for a given dose of air entraining admixture. Interactions with other chemical admixtures may change performance. Early strengths increase. Later strengths decrease. Risk of alkali silica reaction increases (with reactive aggregate). Risk of cracking may increase under certain conditions. Increased reactivity toward supplementary cementing materials. Increase in water requirement. Increase in rate of heat of hydration.
C_3A (increasing) C_4AF (decreasing)	Workability will decrease. Flash set is possible. The rate of heat generation will increase. Early strength increases. Later strengths should not be affected. Cement color may be lighter. Less resistance to sulfates (in a sulfate environment). Risk of delayed ettringite formation may increase (in a suitable environment). Binding capacity of chloride possible.
C_3S (increasing) C_2S (decreasing)	Workability should not be affected. Initial set time will be reduced a small amount. The rate of heat generation will increase. Early strengths are increased. Later strengths may be influenced in either direction depending on relative proportions of C_3S and C_2S. Increase in supplementary cementing material possible.
Cement mortar strength (increases)	Concrete strength will increase (correlation for early strength is fair, for late strength is poor).
Cement set time (increases)	Concrete set time will increase (correlation is poor).
Cement sulfate (ratio of gypsum to hemihydrate decreases)	Risk of false set increases. Strengths may change depending on the optimum sulfate content of the cement and the age at which strengths are measured.
Cement temperature (increases)	Risk of false set increases.
Fineness (increasing)	Workability decreases and water requirement increases. Air content decreases for a given dose of air entraining admixture. Set time decreases. The rate of heat generation will increase. Early strength increases. Later strengths should not be affected. Shrinkage may increase.
Free lime (increasing)	Risk of expansion due to unsoundness increases.
Magnesium (increasing)	Risk of expansion due to unsoundness increases.

(continued)

Table 2.3 Effects of Changes in Cement Characteristics on Concrete Properties (cont.)

	Influence on Concrete (Assuming Only a Change in the Given Component and Nothing Else)
Minor components	Changes in contents of minor components are unlikely to result in observable change in concrete performance; however, F- and P_2O_5 are known to increase setting time.
Particle size distribution (steeper curve)	Water requirement increases. Early strength increases. Porosity may increase. Setting time decreases.

Source: Johansen, V., and Taylor, P. C., "Effect of Cement Characteristics on Concrete Properties," PCA EB 226, Portland Cement Association, Skokie, IL, 2005. With permission.

type of fly ash and the fineness. Silica fume will tend to reduce water requirement at dosages below about 5% but will require the use of chemical admixtures to maintain workability at higher dosages.

- Bleeding rates are generally slower, making mixtures more sensitive to conditions leading to plastic shrinkage cracking. This is especially marked in mixtures containing silica fume where bleeding is effectively stopped.
- Initial hydration is generally slower leading to longer setting times, less heat generation, and lower early strengths, all of which influence the risk of early age cracking and the need to protect the concrete from the environment.
- Longer-term hydration continues longer, potentially leading to greater strengths and improved durability, if there is sufficient water for the reactions to continue.

Aggregates

Aggregates affect the fresh properties of a mixture because the shape, texture, and particle size distribution influence the workability of the mixture. Aggregates are effectively nonreactive and as such do not influence the hydration of the system. The exception to this are lightweight, highly absorbent materials that will provide internal curing to the mixture by allowing water to be removed from their pores. Increasing void space between the aggregate particles will require increasing amounts of paste to fill the voids, thereby leading to a greater potential for drying shrinkage if drying is permitted to start early on.

Water

The largest issue with water in concrete is that the higher the water content in the original mix, the greater the workability. On the other hand, the water–cement ratio will increase, leading to a lower the performance. Conversely, after the mixture has hardened, water must be made available for long-term hydration to proceed.

As long as water is potable it is considered acceptable for use in concrete, although dirty water can still be used if its effects on setting time and early strength are acceptable.

Admixtures

Chemical admixtures are added to concrete to enhance the performance such as providing freeze–thaw resistance, reducing the amount of water required to achieve workability, or changing the set time. Chemical admixtures should not be used to rescue a fundamentally bad mix. Dosages are specific to the product in use, the effect desired, the chemistry of the other materials in the mix, and the temperature. As such, dosages are likely to change frequently to achieve the same mixture performance from batch to batch.

Specialty admixtures are also available that are intended to reduce drying shrinkage, reduce expansion due to alkali silica reaction, and protect reinforcing steel from corrosion.

EFFECTS OF WATER

Hydration of cement and cementitious materials involves converting powders in solution to solid materials. These changes take place over time and can better understood by considering the relative volumes of the various materials at different stages.

Figure 2.7 illustrates the proportions, by volume, of a paste with a water–cement ratio of 0.48 at three stages (Neville 1996):

- Prior to hydration
- 50% hydration
- 100% hydration

If it is assumed that the paste has initial volumes of 60 ml of water and 40 ml of cement, and no outside moisture is added or lost, then the volume of hydration products of the cement is 61.6 ml of solid including 21.6 ml of chemically combined water. Added to this volume there is 24.0 ml of pores that are filled with adsorbed gel water and 7 ml of free water in the capillary pores. The total volume therefore equals 92.6 ml, which is 7.4 ml less than the original volume (Neville 1996).

Based on this model, if the water–cement ratio is less than a critical value of about 0.42 in a sealed system, there will be insufficient water to hydrate all the cement, and the fully hardened paste will consist of cement gel,

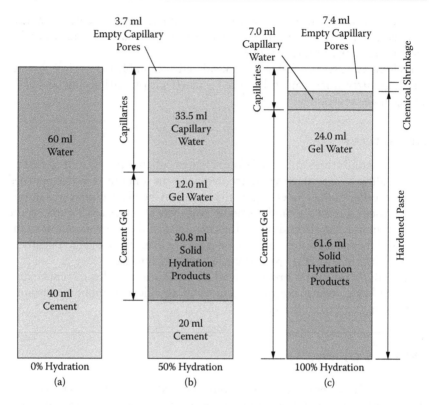

Figure 2.7 An illustration of the relative volumes of materials in a hydrating paste system with a w/c of 0.475. (From Meeks, K. W., and Carino, N. J., "Curing of High-Performance Concrete: Report of the State-of-the-Art," NISTIR 6295, National Institute of Standards and Technology, Gaithersburg, MD, 1999. With permission.)

empty capillaries, and unhydrated cement. If the water–cement ratio is greater than this critical value, all of the cement can hydrate, and the capillary voids will be partially filled with water.

However, if an external supply of water is available to keep the capillary pores saturated, hydration continues until either all the cement has hydrated or all the available space is filled by cement gel. Under this condition the critical water–cement ratio is considered to be about 0.36. This is only valid if water has access to all of the cement. However, at low water–cement ratios the capillaries become disconnected after some hydration thus limiting water access. Therefore, there are likely empty capillary voids in all systems including those cured under water in low w/cm systems. However, if water is allowed to escape, the amount of unhydrated cement and empty voids will markedly increase.

Figure 2.8 An example of a corner break, likely exacerbated by curling.

Although excess water is deleterious during the mixing and placing stages of construction, because it raises w/cm and thereby reduces performance, insufficient water during the first few days of hydration is equally deleterious because hydration slows or stops. This can lead to wasted cement and a mixture that fails to meet performance requirements.

In addition, if water is allowed to escape from a system, it will tend to shrink, with an increasing risk of cracking, particularly at early ages. Differential moisture levels between the top and bottom of a slab may result in warping, in which the corners lift up. Traffic on such a system is likely to increase corner cracking because the slab is unsupported where it has lifted (Figure 2.8).

It is therefore critical that water be continually available to the hydrating system for the first few days and desirable that it be available for a longer period to make the most of the mixture.

TEMPERATURE EFFECTS

Like any chemical reaction, the higher the temperature, the faster hydration proceeds, roughly doubling in rate for every 10°C (18°F) increase. This also means that if the concrete is placed in cold weather, setting time and the rate of strength development will slow significantly, requiring extra protection to prevent shrinkage cracking, and forms must remain in place for longer. Contractors need to be aware of this effect because if forms for an elevated section are stripped based solely on a timetable following a freezing night, it is likely that the concrete will collapse or deflect an unacceptable amount. This is commonly observed in climates that are normally warm but suffer an occasional cold night.

Slabs on grade are particularly sensitive to temperature drops because of the large surface area exposed on one face, leading to differential conditions through the thickness of the slab. Combined with high restraint on the bottom face, these conditions place the system at high risk of curling and cracking. Curling is when the corners or edges of a slab move upward because of tensile stresses set up in the top surface because the top is cooler than the bottom. Self-weight or imposed loads commonly result in corner breaks (Figure 2.8) in curled slabs. A rapid drop in temperature, such as a passing cold front, in a young slab is also at high risk of cracking because temperature shrinkage stresses may be high while strengths are as yet inadequate to carry them.

High temperatures early on may be beneficial when rapid turnaround of the forms is desired, such as in a precast plant. Steam curing provides a cost-effective way to take greatest advantage of the capital intensive formwork because hydration is accelerated and ample moisture is provided to the system. The side effects of this approach are that the microstructure of the hydrated cement paste is coarser, leading to long-term strengths and impermeability values that are not as good as those of a slower hydrating system (Kjellsen et al. 1991).

The other caveat in relation to elevated temperatures is that systems with a certain cement chemistry (high alkalis and C_3A contents) that are exposed to internal temperatures greater than about 65°C or 70°C may be subject to so-called delayed ettringite formation. This phenomenon results in the high temperatures causing ettringite, normally formed in the early stages of hydration, to decompose. At a later stage, months to years, in the presence of copious water the ettringite is redeposited into voids. Such ettringite is expansive and may cause internal cracking (Shimada 2005).

Another system where elevated temperatures may play a role in the performance of a mixture is in mass concrete. In large elements such as pier footings and dam walls, the insulating effect of the large volume of concrete prevents heat generated by the hydration process from escaping the system. The temperature of the interior concrete may rise to temperatures above 130°F while the surface is at, or close to, ambient. The stresses set up by this differential can be sufficient to cause cracking. The temperature profile of the system has to be controlled; normally by precooling the mixture ingredients to reduce the initial temperature, pumping cold water through pipes embedded in the structure, or by wrapping the structure in blankets to prevent the surface from cooling too quickly.

These topics are discussed in more detail in Chapter 3.

SUMMARY

This chapter discussed the hydration process and made it clear that the chemical reaction between portland cement and water is initially fast, but soon slows and continues for a long time as long as water is available. Supplementary cementitious materials generally slow this process but continue for longer. It is therefore critical that to obtain the best performance from the mixture and to obtain the best value from the materials used, the concrete has to be maintained in a condition that allows hydration to continue for as long as possible. This includes preventing drying and keeping the mixture warm, which is the basis of a definition of curing: "action taken to maintain moisture and temperature conditions in a freshly placed cementitious mixture to allow hydraulic cement hydration and (if applicable) pozzolanic reactions to occur so that the potential properties of the mixture may develop" (ACI 2010).

Water must be continuously available. A system that is allowed to dry out, then rewetted will be much harder for water to penetrate, depending on the degree of hydration of the mixture.

Another important consideration is that externally applied curing only affects the surface of the concrete to a depth of about 50 mm (Poole 2006). This means that curing will not significantly affect strength of a structural element such as a column, because the volume of concrete inside the heart of the element is much greater than the so-called covercrete layer. However, this does not mean that curing is not required, because it is that same outer layer that is exposed to environmental attack such as freezing and thawing or carbonation; this same layer is the portion of the concrete that provides protection to the reinforcing steel, and in slabs it is this layer that has to carry the traffic for which it was built. Buildings may not fall down because of poor curing, but the serviceability of all forms of concrete infrastructure will be compromised if curing is not applied effectively. On the other hand, curing can be a very cost-effective means of improving the potential durability of concrete-based elements because it increases the efficiency of the relatively expensive reactive materials that are at the heart of a good concrete mixture.

A relatively new approach to providing moisture to the mixture is to incorporate components into the concrete mixture that serve to deliver moisture to the hydrating cementitious materials from within the mixture; otherwise known as internal curing. This may be achieved by incorporating moist or saturated absorbent material, such as lightweight aggregate sand (LWAS), into the concrete mixture. These porous materials act as internal reservoirs, providing a source of water to replace that consumed by hydration, particularly in systems with a very low water–cement ratio. As the cement hydrates and water is extracted from the capillaries, this extra water will be desorbed from the relatively large pores in the curing agents into the much smaller ones in the cement paste and allow hydration to continue without desiccating the paste. This topic is discussed in more detail in Chapter 4.

The following chapter addresses the questions of how and why curing can be a cost-effective benefit for most concrete elements.

REFERENCES

American Coal Ash Association (ACAA), 2003, "Fly Ash, Facts for Engineers," FWHA-IF-03-019, Federal Highway Administration, Washington, D.C., http://www.fhwa.dot.gov/pavement/fatoc.htm.

American Concrete Institute (ACI), 2010, "ACI Concrete Terminology," Farmington Hills, MI, http://terminology.concrete.org (accessed September 2012).

American Concrete Institute (ACI) Committee 211, 1991, "Standard Practice for Selecting Proportions for Normal, Heavyweight and Mass Concrete," ACI 211.1-91, American Concrete Institute, Farmington Hills, MI.

American Concrete Institute (ACI) Committee 232, 2003, "Use of Fly Ash in Concrete," 232.2R, American Concrete Institute, Farmington Hills, MI.

American Concrete Institute (ACI) Committee 233, 2003, "Slag Cement in Concrete and Mortar," ACI 233R-03, American Concrete Institute, Farmington Hills, MI.

American Concrete Institute (ACI) Committee 234, 2003, "Guide for the Use of Silica Fume in Concrete," ACI 234R-06, American Concrete Institute, Farmington Hills, MI.

Cabrera, J. G., Claisse, P. A., and Hunt, D. N., 1995, "A Statistical Analysis of the Factors Which Contribute to the Corrosion of Steel in Portland Cement and Silica Fume Concrete," *Construction and Building Materials*, vol. 9, no. 2, April, pp. 105–113.

Grove, J., and Taylor, P. C., 2012, "Will More Cement in Your Mixture Hurt You?" 10th International Conference on Concrete Pavements, International Society of Concrete Pavements, Quebec City.

Holland, T. C., 2005, "Silica Fume User's Manual," FHWA-IF-05-016, Federal Highway Administration, Washington, D.C.

Hooton, R. D., 1993, "Influence of Silica Fume Replacement of Cement on Physical Properties and Resistance to Sulfate Attack, Freezing and Thawing, and Alkali-Silica Reactivity," *ACI Materials Journal*, vol. 90, no. 2, pp. 143–151.

Johansen, V., and Taylor, P. C., 2005, "Effect of Cement Characteristics on Concrete Properties," PCA EB 226, Portland Cement Association, Skokie, IL.

Kjellsen, K. O., Detwiler, R. J., and Gjorv, O. E., 1991, "Development of Microstructures in Plain Cement Pastes Hydrated at Different Temperatures," *Cement and Concrete Research*, vol. 21, pp. 179–189.

Koehler, E. P., and Fowler, D. W., 2010, "ICAR Mixture Proportioning Procedure for Self-Consolidating Concrete," International Center for Aggregates Research, University of Texas at Austin, ICAR Project 108: Aggregates in Self-Consolidating Concrete.

Kosmatka, S. H., and Wilson, M. L., 2011, *Design and Control of Concrete Mixtures*, 15th ed., EB001, Portland Cement Association, Skokie, IL.

Marusin, S. L., 1989, "Influence of Length of Moist Curing Time on Weight Change Behavior and Chloride Ion Permeability of Concrete Containing Silica Fume," Proceedings of the Third International Conference, Fly Ash, Silica Fume, Slag, and Natural Pozzolans in Concrete, vol. 2, ACI SP-114, V. M. Malhotra, ed., American Concrete Institute, Farmington Hills, MI, pp. 929–944.

Meeks, K. W., and Carino, N. J., 1999, "Curing of High-Performance Concrete: Report of the State-of-the-Art," NISTIR 6295, National Institute of Standards and Technology, Gaithersburg, MD.

Neville, A. M., 1996, *Properties of Concrete*, 4th ed., John Wiley & Sons, New York, NY, pp. 25–37.

Poole, T. S., 2006, "Curing Portland Cement Concrete Pavements, Volume II," FHWA-HRT-05-038, Federal Highway Administration, McLean, VA.

Shilstone, J. M., Sr., 1990, "Concrete Mixture Optimization," *Concrete International*, vol. 12, no. 6, pp. 33–39.

Shimada, Y. E., 2005, "Chemical Path of Ettringite Formation in Heat Cured Mortar and Its Relationship to Expansion," Ph.D. thesis, Northwestern University, Evanston, IL.

Siddique, R., and Khan, M. I., 2011, *Supplementary Cementing Materials*, Springer, Berlin.

Sutter, L., 2008, personal communication.

Taylor, P. C., Johansen, V. C., Graf, L. A., Kozikowski, R. L., Zemajtis, J. Z., and Ferraris, C. F., 2006a, "Identifying Incompatible Combinations of Concrete Materials, Volume I—Final Report," Report HRT-06-079, Federal Highway Administration, Washington, D.C.

Taylor, P. C., Kosmatka, S., and Voigt, J., eds., 2006b, "Integrating Materials and Construction Practices for Concrete Pavements," HIF-07-004, Ames, IA, National Concrete Pavement Technology Center, Iowa State University, Federal Highway Administration.

Thomas, M., 2013, *Fly Ash*, Taylor & Francis, Boca Raton, FL.

Yurdakul, E., 2010, "Optimizing Concrete Mixtures with Minimum Cement Content for Performance and Sustainability," Master of Science thesis, Iowa State University, Ames, IA.

Chapter 3

Benefits of curing on concrete performance

INTRODUCTION

This chapter describes the common causes of failure in concrete structures and pavements, and discusses how the risk of their occurrence can be reduced by curing (that is, providing adequate warmth and moisture to a hydrating concrete system) particularly at very early ages.

CRACKING

Fundamentals

Concrete almost always cracks, because it is a brittle, heterogeneous material, weak in tension that undergoes significant movements due to internal temperature moisture and temperature changes. The challenge is to limit the width and extent of cracks so that the concrete is able to perform as intended. To do this, practitioners need to understand the mechanics of crack development.

Stress, strain, and cracking

Whenever a force is applied to a fixed object, it will deflect, albeit a small amount. Under that force, the material in the element will become stressed, with the magnitude of the stress being governed by the effective area of material resisting the force.

Stress can also be induced by internal movements such as restrained shrinkage or differential expansion within the system. A system that is not restrained will simply move, while restraint will set up stresses as the object is deformed. The restraint may be from external connections such as friction on the ground, or from internal reinforcing steel.

As movement increases in restrained systems, the stress on the material also increases. The magnitude of the stress is governed by the magnitude of the strain (movement) and the stiffness of the material. Stress will rise faster in a stiff material than in a more flexible material. For example, a rubber band exerts less force than a similar sized piece of steel when stretched the same amount, because rubber is less stiff than steel.

The stress–strain relationship is a property of the material and the slope of stress over strain is a measure of the stiffness. At low loads the stress–strain relationship for concrete is linear, but at higher loads it will start to behave nonlinearly as internal damage or microcracking starts to accumulate.

When the stresses at any point in the element are greater than the local strength of the material, a crack will form. Stresses are then redistributed because the amount of material available to carry them has changed. This redistribution in the heterogeneous system may also lead to multiple micro-cracks in different locations. As stresses increase, a macrocrack will emerge and propagate, eventually leading to failure.

Finally, cracking in any material will tend to occur in the mode that is the weakest for that material. For example, mortar and concrete are strong in compression and weak in tension and shear. As predicted in a Mohr's circle diagram, the shear stresses at 45° to the vertical become the critical factor, even when vertical loads are applied to an element. This is commonly observed in mortar cube compression tests where the shape of a properly tested sample is a square cone as shown in Figure 3.1, because failure has occurred in shear at 45° to the direction of the force.

Figure 3.1 Photograph of a mortar cube loaded in uniaxial compression. The planes of failure are as predicted by Mohr's circle.

Therefore, in summary, cracking occurs when local stresses induced by applied loads or restrained movement are greater than the local strength of the material carrying the load.

Restraint of volume change

Volume change occurs within cementitious materials because

- The volume of hydration products is less than that of the original raw materials (autogenous shrinkage).
- Moisture is lost from the system, either before setting (plastic shrinkage), after setting (drying shrinkage), or due to internal desiccation.
- Temperature changes.

These effects are additive, develop at different rates, and may have different magnitudes. In many cases, addressing one of them, even if it is not the largest at any given time, will help to reduce the risk of cracking. Alternatively, delaying them until sufficient strength is developed in the mixture will also help to reduce risk. This approach does have to be balanced with the fact that with increasing strength in concrete, stiffness also increases and often at a relatively greater rate, effectively leading to a possible greater risk of cracking. In addition, rapid drying is likely to lead to nonuniform moisture profiles, which in turn will cause potentially high stress fields near the surface.

Under conditions of no restraint, no stress is imposed; the body expands or contracts freely. If there is restraint, stresses develop in proportion to the stiffness of the material and the degree of restraint. Restraint may be external, for example, due to bond with the substrate material, or internal, for example, if the outer concrete shrinks while the core does not.

In practical cases, internal and external restraints almost always coexist. External restraint can arise from support at the ends or intermediate supports of beams and slabs if the deformation is restricted by adjacent members. Slabs on grade or foundations cast directly against soil experience continuous restraint, with the degree of restraint dependent on the shear resistance of the soil. Abrupt changes of cross-section within a structural element will also lead to restraint. Aggregate particles within concrete can impose internal restraint due to differences in the stiffness of the aggregate and paste.

Mechanical properties of concrete

Concrete cracks when and where the tensile stress exceeds the tensile strength. Thus, high tensile strength provides the greatest resistance to cracking. Stresses are proportional to the modulus of elasticity of the concrete, therefore high modulus of elasticity, which tends to correspond with high strength, contributes to cracking by increasing the stress. The development of the mechanical

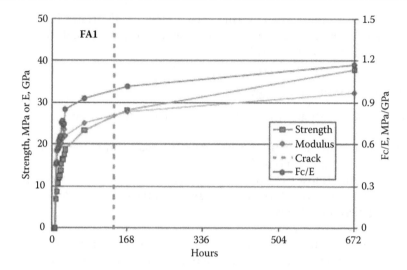

Figure 3.2 Plots of measured strength and modulus for a single mixture, compared with when cracking occurred in a ring shrinkage test. Note that modulus increases more rapidly than strength initially (even allowing for the different scales).

properties, particularly within the first 24 hours, is critical to the prevention of cracking. Unfortunately modulus of elasticity tends to develop faster than strength in the first few hours after setting (Figure 3.2; Taylor 2006).

Because concrete is a heterogeneous material, the distribution of stress may not be uniform, and likewise the local strengths are also not uniform. For instance, the strength of the paste is not the same as that of the aggregate or the interfacial zone between them. This results in the formation of a zone of microcracks that are not necessarily connected until stresses increase; some microcracks merge and others close, and a single crack starts to grow. It is also not uncommon to see cracks jump around aggregate particles as the stress field redistributes or microcracks that are oriented in all directions at low loads (Figure 3.3).

Shrinkage

Plastic shrinkage

If the rate of evaporation is higher than that of bleeding, the air–water interface will move below the surface of the concrete matrix. Consequently, the capillary pores will be partially filled, and menisci will form at the interface between air and water in the capillaries (Poole 2006). The direction of the forces set up by the surface tension at the interface have a component normal to the face of the capillary, therefore closing the capillary on drying, resulting in shrinkage of the concrete (Chatterji 1982; Figure 3.4).

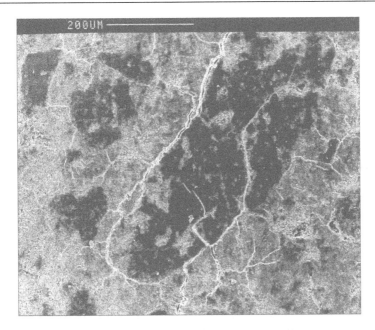

Figure 3.3 Micrograph illustrating randomly oriented cracks in a damage zone. Principal crack growth direction is top to bottom in the image.

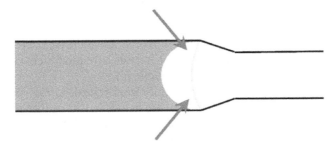

Figure 3.4 Sketch illustrating the closing forces set up by the water meniscus as a capillary dries out.

This shrinkage can occur both when the concrete is still "plastic" and also after final set. If it occurs before setting, there is sufficient resistance to displacement that cracking may occur, known as plastic shrinkage cracking.

Plastic shrinkage cracking, therefore, can be prevented by preventing the evaporation of water from the surface. Concrete exposed to rapid drying caused by wind, elevated temperatures, or low humidity is at increased risk of plastic shrinkage cracking (Holt 2000).

Plastic shrinkage cracks normally appear on the surface of concrete during or soon after finishing. They are usually parallel to one another, discontinuous, and often perpendicular to the direction of the prevailing wind (Marais and Perrie 1993); however, they may also form random patterns. They range from a few inches to many feet in length and are spaced from a few inches to as much as 10 feet (3 m) apart, with depths ranging up to the thickness of the section (American Concrete Institute [ACI] Committee 224 1993).

Increasing the cementitious materials content increases the likelihood of plastic shrinkage cracking because of reduced bleeding. For the same reasons, the use of silica fume makes the concrete particularly vulnerable to plastic shrinkage cracking. Increased slump increases the risk of plastic shrinkage cracking because it is likely tied to additional water in the mixture. Delays in setting increase the tendency for plastic shrinkage cracking when the rate of evaporation is high, because the concrete is plastic for a longer time. Such delays may be caused by cool weather, cool subgrades, low cement content, pozzolans or slags, retarders, and most water reducers (National Ready Mixed Concrete Association [NRMCA] 1992).

NRMCA (1992) recommended the following precautions to minimize plastic shrinkage cracking:

- Have sufficient personnel, equipment, and supplies on hand to place and finish the concrete promptly. Cover the concrete with wet burlap, polyethylene sheeting, or building paper between finishing operations to prevent drying.
- Start curing the concrete as soon as possible.
- Dampen the subgrade, formwork, and reinforcement before placing concrete.
- Vapor barriers directly under a slab on grade increase the risk of plastic shrinkage cracking. If a vapor barrier is required, cover it with a 50 mm layer of compactible drainable fill.
- In very hot, dry weather, use evaporation retarders, fog sprays, temporary windbreaks, and sun shades as needed. Consider placing the concrete in late afternoon, early evening, or at night.
- Synthetic fibers may help to control plastic shrinkage cracking.
- Make the concrete set faster.

A related problem is plastic settlement cracking, which occurs due to incomplete consolidation of the concrete or to bleeding. Aggregate particles tend to sink and water rises to the surface. The movement may result in cracking appearing over reinforcing steel or other restraint. The cracking normally only penetrates as far as the steel. Revibration before the concrete sets will remedy this problem.

Drying shrinkage

Similarly to plastic shrinkage, drying shrinkage is a result of moisture loss from the capillaries, except that it occurs after final set. Once again, the

Figure 3.5 Star crack due to drying shrinkage.

forces on the capillaries result in shrinkage of the paste. The difference is that the form of cracking, if any, will be different. Extensive microcracking can occur even with mild drying. As drying continues, some microcracks may coalesce to form visible drying shrinkage cracks (Chatterji 1982).

The magnitude of drying shrinkage is normally about double that of thermal shrinkage, but it occurs over a much longer period. Drying shrinkage can continue for many years, although laboratory tests show that approximately 80% of drying shrinkage takes place within the first 3 months (Holt 2001). Cracking due to drying shrinkage also appears to be related to the rate of drying, regardless of the strength of the system (Tongaroonsri and Tangtermsirikul 2009). This is likely tied to increased stresses set up by differentials through the depth of a concrete element with rapid drying.

Concrete flatwork is most vulnerable to drying shrinkage cracking because of its high surface-to-volume ratio. A common form of drying cracking is a so-called star crack in slabs on grade (Figure 3.5). Restrained elements of any form are at high risk of drying shrinkage cracking at the time that forms are removed. This is because the rate of drying is high, leading to large differentials, at the same time that stiffness is relatively high and strength is still increasing (Altoubat and Lange 2001). It is therefore essential to ensure that concrete is prevented from drying for some time after the forms are removed until strengths are sufficient to resist drying stresses.

To minimize drying shrinkage cracking:

- Continue curing as long as needed; for example, a bridge deck might require 7 days of moist curing.

- On removal of forms, protect the concrete from drying, for example by applying a curing compound.
- Consider using a shrinkage-reducing admixture.

Autogenous shrinkage and self-desiccation

In standard testing of concrete according to ASTM C157, drying shrinkage is measured after the specimens have been moist cured for 28 days. However, as Holt (2001) points out, this practice is erroneous because the early-age shrinkage can equal or even exceed the long-term shrinkage.

The volume of the hydration products of cement is less than that of the unhydrated cement and water from which they form. This is most noticeable in concrete in which the water–cement ratio is less than about 0.42. In addition hydration of cement in low water–cement mixes after setting may self-desiccate the microstructure, again causing water menisci to form and further shrinkage (Chatterji 1982). Together these mechanisms are known as chemical shrinkage.

Chemical shrinkage is defined as the total volume change due to hydration and desiccation of cement. However, when exterior dimensional measurements are taken, some of the deformation is not detected because of the formation of voids within the matrix (Kosmatka and Wilson 2011). Autogenous shrinkage is defined as the measurable volume change that occurs. These mechanisms that take place within the first 24 hours of placement are of concern because at these early ages the concrete has limited strength.

Concrete mix proportions and ingredients will have the most significant influence. The following can be considered to reduce the amount of autogenous shrinkage:

- Increase water–cement ratio (note that this may increase drying shrinkage).
- Use higher proportions of aggregate or water reducing admixtures to reduce the amount of cement paste available to shrink.
- Use a cement with lower C_3A.
- Avoid conditions that will delay setting time.

Thermal effects

Traditionally, thermal cracking is associated with mass concrete: dams, foundations, and large structural members in which the heat of hydration is not readily dissipated. Early cracking in massive concrete members is due predominately to thermal stresses (Rostásy et al. 1998).

Thermal stresses occur when thermal strains (expansion on heating, contraction on cooling) are restrained, either externally or internally. Typically problems arise when forms are removed, leading to cooling at the surface while the mixture is still hot internally. As a result of this thermal contraction, the surface will be in tension and the interior in compression, and if the tensile stresses on the surface exceed the tensile strength, then the concrete will crack. The practices necessary to control cracking in this type of application are well established and are described in ACI 207 (1996) and discussed later.

The use of high-performance concrete presents more of a challenge with regard to thermal cracking. Generally, high-performance concrete contains relatively high cementitious materials contents, leading to greater heat being generated at earlier ages and high thermal gradients even in relatively thin sections. These concretes are also more vulnerable to autogenous shrinkage and self-desiccation, both of which will further increase the tensile stresses and thus the risk of cracking. At the same time, some of the traditional practices used to control thermal cracking in mass concrete do not lend themselves to high-performance applications.

Slabs on grade are particularly prone to thermal-related cracking because the concrete changes from a liquid to a solid at about the same time it is at its maximum temperature due to ambient conditions and the heat of hydration. Concrete placed in the morning typically undergoes rapid cooling in the evening while it is a solid but without much strength. The large surface-to-volume ratio means that changes in ambient temperatures will have a marked influence on the stresses in the slab. Thermal cracking in slabs is typically only a concern for the first 24 to 48 hours when the internal temperature of the mixture is high and falling.

Another factor to be considered with respect to temperature is the differential imposed when a surface is suddenly exposed to much cooler external conditions, such as when forms are removed, or if cold water is sprayed on a hot surface. ACI 305 recommends that the maximum difference between interior and surface temperatures should be no less that 10°C and the Army Corps of Engineers limit is 13°F. Data collected by Poole (2006) indicates that an evaporation rate of 1.4 kg/m²/h imposed on a sample at 44°C can induce such a differential (Figure 3.6).

Reducing the risk

Materials

To reduce the risk of temperature-related cracking, materials selection and proportioning should be based on addressing the following:

- Development of strength, stiffness, and creep resistance. In general, the characteristics of concrete that contribute to high strength also result in high stiffness, and high stiffness also generally coincides with a reduced tendency to creep.

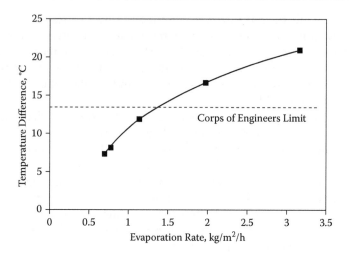

Figure 3.6 Plot showing relationship between evaporation rate and temperature differential in samples at 44°C. (Image from FHWA.)

- Tendency to shrink. The tendency to shrink is associated with the paste fraction of the concrete and the fineness of the powder. Thus minimizing the paste content by an appropriate aggregate particle size grading is the best way to minimize shrinkage. Water-reducing admixtures also can be used to reduce the paste content while maintaining workability. Shrinkage-reducing admixtures may also be used to reduce the drying shrinkage of a mixture.
- Heat of hydration. Heat is generated by the reaction between water and cement and, to a lesser extent, supplementary cementitious materials. Thus, the type and quantity of cementitious materials are important factors in controlling heat of hydration and the resulting thermal stresses.

Mix design and proportioning

The following practical measures are recommended when selecting materials and proportioning mixes to minimize cracking in concrete (ACI Committee 224 1993).

- Minimize the paste content in the mix, within other constraints imposed by the need for strength or potential durability.
- Minimize the water content.
- Use as large an aggregate as possible within the limits set by section thickness and reinforcing spacing.

- Entrain air (about 4% to 6% by volume) even if it is not required for frost resistance, unless the concrete is to receive a hard-troweled finish.
- Do not specify a higher strength than necessary. Consider specifying a 56- or 90-day strength rather than a 28-day strength to reduce total cementitious content.
- Select a cement with a low alkali content and optimal sulfate content and that is not too fine.

Curing

As noted earlier, cracking risk can be significantly reduced if water is kept in the system for as long as possible to allow hydration to proceed and mechanical properties to develop before drying shrinkage stresses are induced (Nabil et al. 2011). In addition, protecting the young concrete from thermal fluctuations will help to reduce risk of damage. Methods of how this may be achieved are discussed in a later chapter.

DURABILITY

Introduction

Frequently the most important characteristic of concrete is not strength but durability (Detwiler 2005). Strength is sometimes specified as a surrogate for durability because it is easy to test, and because it is indirectly related to durability in many instances; some of the actions needed to improve durability also result in increasing strength. However, the two are not the same (Figure 3.7), and in some cases, the measures that produce high strength are detrimental to durability. For example, concrete subject to cycles of freezing and thawing should be air entrained; however, air entrainment usually reduces the strength. Another example is that increased strength is often obtained by increasing the cement content. However, increased cement content also increases the risk of shrinkage-related cracking. It is important to distinguish between strength and durability, and to specify as directly as possible the properties required.

The various mechanisms by which concrete can be attacked are discussed in this chapter. In all cases, suggestions are made on how to minimize the risk or rate of attack when designing or building a structure likely to be subjected to the mechanism being discussed. It should be noted that concrete in a given environment might be subjected to more than one form of attack. If so, the various forms of attack must be considered together and the concrete designed to withstand all of them simultaneously. Concrete with an adequate cement content and a low water–cementitious materials ratio, which has been mixed, placed, consolidated, finished, and cured according to good concrete

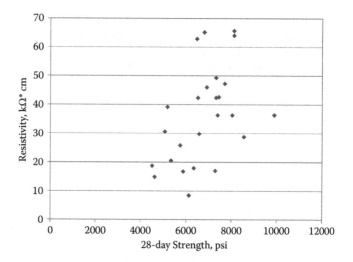

Figure 3.7 Plot of concrete resistivity (a measure of potential durability) against strength for mixtures containing a variety of cementitious binders. (After Taylor, P.C. et al., "Development of Performance Properties of Ternary Mixtures: Field Demonstrations and Project Summary," DTFH61-06-H-00011 Work Plans 12 and 19, Federal Highway Administration, 2012.)

practice, is inherently more durable than a concrete that deviates from this ideal in any way. In the following sections, this principle is assumed and not repeated unless particular attention must be paid to some detail.

Chemical mechanisms

The fundamental mechanisms behind chemical attack of concrete are relatively limited but complex. Cement comprises a number of compounds, primarily comprising combinations of lime (CaO), silica (SiO_2), and alumina (Al_2O_3). In addition, concrete contains aggregates and in some cases reinforcing steel. The sensitivity of these materials includes

- Calcium-based compounds tend to be soluble in acids and in soft water, and will react with carbon dioxide.
- Calcium alumina compounds will react with sulfates to form expansive by-products that disrupt the matrix.
- Calcareous aggregates are more prone to abrasion and polishing, and are subject to acid attack.
- Siliceous aggregates may be subject to alkali–silica reaction.
- Reinforcing steel can corrode if the concrete pore solution pH drops sufficiently or if chlorides are present.

These mechanisms are described in more detail in the following sections.

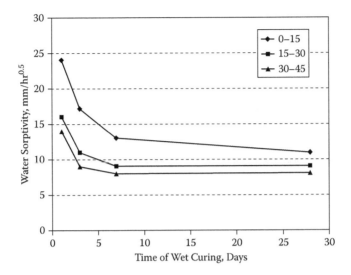

Figure 3.8 Plot illustrating the effect of wet curing at various depths (in millimeters) on sorptivity over time. (After Ballim, Y. et al., "Assessment and Control of Concrete Durability," Concrete Meets the Challenge, National Convention of the CSSA, September, 1994.)

It should be noted that in all cases of chemical attack, the presence of water is necessary as a reactant or to transport the aggressive ions. It is therefore logical that reducing concrete permeability will significantly reduce the risk or rate of damage from all forms of chemical attack. For this reason, measures such as reducing the water–cementitious materials ratio, providing extended moist curing (Figure 3.8), and the judicious use of supplementary cementitious materials are effective in improving the resistance of concrete to a variety of chemical attack mechanisms.

Water is readily easily transported through capillaries, while permeability of the solid matrix is very low. The overall permeability, then, is strongly tied to the continuity of the capillaries in the system. The higher the water–cementitious materials ratio (w/cm), the greater the probability that capillaries will percolate (be connected through the width of the section). Alternatively, the higher the w/cm, the longer it will take for sufficient hydration products to form to lead to capillary discontinuity, as illustrated in Table 3.1. This means that higher w/cm mixtures have to be cured for longer to provide protection to the system.

External sulfate attack

Sulfate ions are sometimes found in soil and groundwater, and are also present in seawater, some industrial environments, and sewers. Sulfate ions attack calcium hydroxide (CH, a product of the hydration of the C_3S and

Table 3.1 Time for Capillaries to Become Discontinuous

Water–Cement Ratio	Time
0.40	3 days
0.45	7 days
0.50	28 days
0.60	6 months
0.70	1 year
0.80	Never

Source: Detwiler, K. J., and Taylor, P. C., "Specifier's Guide to Durable Concrete," Portland Cement Association, Skokie, IL, PCA EB 221, 2005. With permission.

C_2S in cement) along with the hydration products of C_3A, forming ettringite in an expansive reaction (Mehta 1986).

Sulfate and monosulfoaluminate (a by-product of cement hydration) also react to form ettringite (Gollop and Taylor 1992). Since additional calcium ion is needed for this reaction, some of the calcium hydroxide dissolves. Once all of the monosulfoaluminate has reacted, the sulfate reacts with calcium ions (from the CH and C-S-H) and water to form gypsum. These reactions usually manifest in the form of loss of cohesion and strength as well as expansion-related cracking. Sulfate attack can lead to loss of strength, expansion, spalling of surface layers, and ultimately disintegration (Detwiler 2005). Thus, sulfate attack produces both ettringite and gypsum while it dissolves calcium hydroxide and decalcifies the calcium silicate hydrate.

The degree of distress depends on the cations involved, with magnesium sulfate attack being the most aggressive and calcium sulfate the least aggressive.

Several approaches can be used to minimize the effects of sulfate attack (Neville 1996):

- Reduce the C_3A content of the cement by specifying cements with low C_3A contents. Note that this action may slow reactions but may not be a guarantee of protection in very aggressive environments.
- Limit the CH content of the concrete by using a supplementary cementitious material to form additional C-S-H.
- Limit the ability of the sulfates to enter the concrete in the first place; including by minimizing the water–cementitious materials ratio and providing good curing.
- Consider providing a waterproof barrier on the surface of the concrete in very severe sulfate exposures.

Internal sulfate attack

Sulfate attack can also occur when the sulfate is supplied internally. ACI Committee 221 (2001) reports the presence of sulfates in a variety of aggregate types. The most common form of sulfate in aggregates is gypsum, which occurs as a coating on sand and gravel, as a component of some sedimentary rock, or in weathered slags. Aggregates made from recycled building materials may contain sulfates in the form of contamination from plaster or gypsum wallboard.

Internal sulfate attack can be prevented by prohibiting the use of aggregates containing gypsum or other sulfate-bearing materials. If petrographic examination of the aggregate shows that the gypsum is concentrated in a particular size fraction, that size fraction can be sieved out and discarded. Sulfate-resistant cements are not adequate to resist internal sulfate attack over the long term.

Delayed ettringite formation

Another form of internal sulfate attack is delayed ettringite formation (DEF). Ettringite is a normal byproduct of cement hydration in the first few hours after mixing. Since ettringite is not stable at elevated temperatures, monosulfoaluminate forms instead when concrete hydrates at high temperatures, even when sufficient sulfate is present. Years later, ettringite re-forms from the monosulfoaluminate in a reaction that causes the paste to expand (Figure 3.9; Famy and Taylor 2001). An abundant supply of water is necessary for the formation of ettringite because each mole of ettringite contains 32 moles of water.

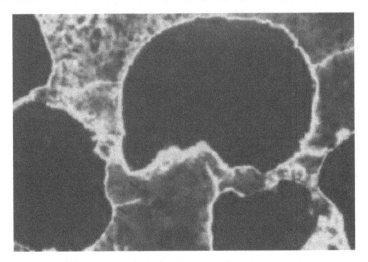

Figure 3.9 Micrograph showing that paste has expanded (mottled), leaving voids (light) around the aggregate (dark) that has not expanded in DEF-affected mortar. (From Detwiler, R. J., and Taylor, P. C., "Specifier's Guide to Durable Concrete," Portland Cement Association, Skokie, IL, PCA EB 221, 2005. PCA image 13655. With permission.)

For mortar or concrete that has been cured at elevated temperatures, the risk of expansion is related to cement composition. Cements with the greatest susceptibility contain high C_3A, C_3S, Na_2O_{eq}, MgO, and fineness (Tracy et al. 2004). DEF is a complex phenomenon, which depends on the concrete temperature during curing as well as on both physical and chemical properties of the system (Shimada 2005). Several competing factors are at work, including the amount of ettringite that decomposes, the rate at which ettringite re-forms, the rate at which stress can be relieved through reprecipitation in larger spaces (Ostwald ripening), and the strength of the system to resist expansive stresses. The DEF-related expansion of a system will therefore be controlled by the curing temperature, the environment in which it exists, and the chemistry of the system.

Miller and Conway (2000) found that slag cement in dosages as low as 5% is effective in reducing expansions due to delayed ettringite formation. Other supplementary cementitious materials (SCMs) may also help reduce the risk of delayed ettringite formation.

Because of the risk of delayed ettringite formation, as well as the deleterious effects of elevated temperature on permeability (Detwiler and Taylor 2005), concrete temperatures above 70°C (158°F) should be avoided in the first few days.

The heat of hydration of a mixture may raise the internal temperature of concrete above the desired limits, even if no additional energy is applied to the concrete. Concrete elements with a minimum dimension of more than about 1 m (3.3 feet) should be monitored and precautions taken against unacceptably high temperatures occurring due to this phenomenon.

Well-cured systems that limit the ingress of water will be at lower risk of DEF-related cracking.

Salt crystallization

Salt crystallization is a physical attack by salts transported into the concrete by water through capillary action and diffusion. When pore water later evaporates from exposed surfaces, or freezes in cold weather, the salt concentrates until it crystallizes (Figure 3.10; Haynes et al. 1996). Changes in ambient temperature and relative humidity also cause some salts to undergo cycles of dissolution and crystallization. When crystallization is accompanied by volumetric expansion, repeated cycles can cause incremental deterioration of concrete similar to that caused by cycles of freezing and thawing. Damage due to salt crystallization can occur with a variety of salts.

An effective way to prevent damage by salt crystallization is to prevent the salt-laden water from entering the concrete and wicking up to an exposed face. Where the climate is arid but the local groundwater table is near the surface, the best way to prevent damage is to provide a barrier on

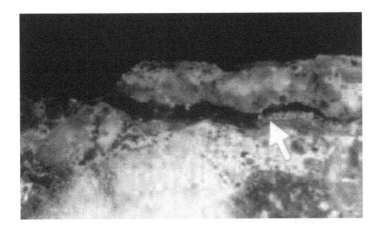

Figure 3.10 Micrograph showing salt deposits in a crack near the surface of a damaged concrete. (From Detwiler, R. J., and Taylor, P. C., "Specifier's Guide to Durable Concrete," Portland Cement Association, Skokie, IL, PCA EB 221, 2005. PCA image 13647. With permission.)

the upstream face. Coating the concrete surface or placing plastic sheeting beneath slabs helps keep the salt solution out.

Any measure that reduces the permeability of the concrete including a low water–cement ratio, use of supplementary cementitious materials, and good curing also reduces the vulnerability of the concrete to damage.

Corrosion of reinforcement

Corrosion of steel is an electrochemical process. For it to take place, all of the elements of a corrosion cell must be present: an anode, a cathode, an electrolyte, and an electrical connection. At the anode (the part that corrodes), the net reaction is

$$2Fe + O_2 + 2H_2O \rightarrow 2Fe(OH)_2$$

$2Fe(OH)_2$ combines with additional oxygen and water to form rust, $Fe_2O_3 \cdot nH_2O$. Figure 3.11 illustrates the various corrosion products that may form and shows that the volume of the final corrosion product may be more than six times the volume of the original iron.

Under normal circumstances, that is, in a concrete where the pH of the pore solution is high, a passive layer of oxides forms on the surface of steel embedded in the concrete. This film protects it from further corrosion. However, the passive layer may break down under certain conditions, including the presence of chlorides, or a local reduction of the pH of the pore solution due to carbonation of the system.

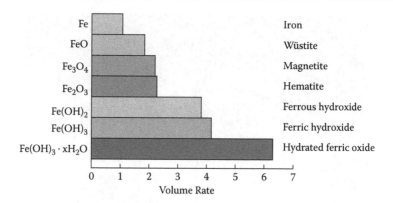

Figure 3.11 Illustration of the relative volumes of various oxides of iron formed when it corrodes. (From Detwiler, R. J., and Taylor, P. C., "Specifier's Guide to Durable Concrete," Portland Cement Association, Skokie, IL, PCA EB 221, 2005. PCA image 13647. With permission.)

Various factors affect the rate of corrosion of steel. These include the following:

- Availability of water and oxygen.
- Reduction of the pH. The protective film is stable above a pH of about 12.5 and corrosion rates are consequently very low.
- Presence of chlorides, which act as catalysts to accelerate corrosion even at high pH.
- Elevated temperature. The rate of corrosion for a given concentration of oxygen increases with increasing temperature.
- Increased electrical resistivity of the concrete. Concrete with a high electrical resistivity will reduce the rate of corrosion by reducing the rate of ion transport.
- Transport properties of the concrete. The diffusivity of the concrete to ions and permeability to fluids affects the rate of corrosion.

Of these factors, the most easily controlled are those pertaining to the properties of the concrete. Keeping the reactants—oxygen, water, and chlorides—away from the steel is essential. A high-quality concrete with sufficient concrete cover over the reinforcing steel and cracks limited to a maximum width of 0.2 mm (0.008 inches) will provide good protection for the steel. Cements with high C_3A contents or slag cement are effective in binding chlorides as they pass through the mixture, thus reducing the rate of penetration. Where stray currents are expected, the electrical resistivity of the concrete should be addressed explicitly.

In bridge decks and parking garages, corrosion-inhibiting admixtures may be used to provide additional protection against corrosion. However, no admixture should be used as a substitute for good concrete practice.

Judicious use of one or more supplementary cementing materials, combined with extended moist curing, can effectively reduce the electrical conductivity and fluid permeability.

Carbon dioxide

Concrete may undergo carbonation due to the action of CO_2 from either the atmosphere or carbonated water. The CO_2 reacts with calcium hydroxide in the concrete to form calcium carbonate, reducing the pH and so accelerating corrosion of reinforcing steel. Carbonation does not normally affect mechanical performance of a mixture, except in extreme cases where C-S-H may decompose to maintain equilibrium. This will only occur in poor-quality systems and other factors are likely to lead to failure first.

A marginally beneficial effect of carbonation is that the carbonate product will tend to densify the surface so slightly reducing permeability.

Measurement of the depth of carbonation provides an indication of the permeability of the system. This can be accomplished by spraying a freshly exposed surface with a phenolphthalein solution or by examination under a petrographic microscope.

The rate of ingress of carbonation into the concrete is most rapid at intermediate relative humidities. Since the reaction involves the dissolution of CO_2 in water, some moisture is needed. However, in very wet concrete the transport of CO_2 is slow because its solubility in water is low.

The rate of carbonation attack is best reduced by ensuring that the concrete is sufficiently impermeable by use of supplementary cementitious materials, low w/cm, and effective curing.

Aggressive waters and other chemicals

Pure, soft, or acidic waters can dissolve the calcium-based components of hydrated cement paste, particularly the calcium hydroxide. The dissolved material leaches out, leaving voids, which results in increased porosity and permeability. The increased permeability accelerates the rate of all forms of deterioration, as water and other harmful materials can more readily penetrate the concrete. In extreme cases, C-S-H may also decompose. Because calcium hydroxide is more soluble in cold water, leaching is more rapid at low temperatures (Mindess and Young 1981).

Acids will react with the constituents of portland cement paste and calcareous aggregates to form soluble salts of calcium, which can leach out of the concrete (Mehta 1986).

Similar to other distress mechanisms, reduced permeability will help in reducing the rate of attack.

Seawater

Eleven ions make up more than 99.99% of the dissolved material in sea-water, thus all of the necessary ingredients are present for at least three different deterioration mechanisms:

- Corrosion (chlorides, oxygen, water)
- Sulfate attack (sulfates, water)
- Magnesium ion substitution (magnesium, water)

Several additional deterioration mechanisms may also come into play:

- Alkali–aggregate reaction (if reactive aggregates are used)
- Abrasion/erosion (due to wave action, sand, gravel, icebergs, ships)
- Frost damage
- Ship impact
- Carbonation
- Salt crystallization

When several deterioration mechanisms take place in the same concrete, they generally interact so that the net effect is greater than the sum of individual effects. For example, cracking due to corrosion will allow carbon dioxide, sulfate, and magnesium to penetrate more readily into the concrete to attack the interior as well as the surface. A notable exception is the effect of chloride ions on sulfate attack. Despite the high sulfate content of seawater, sulfate attack in marine concrete does not result in expansive formation of ettringite. Instead, deterioration takes the form of erosion or loss of the solid constituents (Mehta 1986). This is likely because ettringite expansion is suppressed in environments where OH^- ions are essentially replaced by Cl^- ions.

Deterioration mechanisms vary depending on whether the concrete is located in the atmospheric zone (always above the water), the tidal or splash zone, or the submerged zone (always below the water). The splash zone is the most severe exposure because both oxygen and seawater are in abundant supply; cycles of wetting and drying concentrate the salts present; cycles of freezing and thawing take place when the concrete is fully saturated; and the action of waves, floating objects, and sand create the most severe conditions of abrasion and erosion.

Impermeable concrete will be at lower risk than more open systems.

Alkali–silica reaction

The basic mechanism of alkali–silica reaction is described as follows:

- Alkali ions (Na^- and K^+) from the cement, mixing water, or environment increase the equilibrium concentration of OH^- ions in the pore solution.
- The OH^- ions attack susceptible siliceous minerals in the aggregate, forming a hygroscopic gel.
- The gel attracts water from the surrounding pore solution, swelling and causing cracking, which emanates from the aggregate particle (Figure 3.12).

The necessary conditions for alkali–silica reaction to take place are a reactive mineral in the aggregate, a sufficiently high concentration of alkalis in the concrete pore solution, and a supply of water.

Ideally, aggregates susceptible to alkali–silica reaction should not be used in concrete. However, in some locations these are the only aggregates available. Importing nonreactive aggregates may not be cost-effective. Where reactive aggregates must be used, sufficient quantities of appropriate supplementary cementitious materials should be incorporated into the concrete. The use of low-alkali cement may not be sufficient to control the expansion of all aggregates and is not appropriate where alkalis can migrate into the concrete from

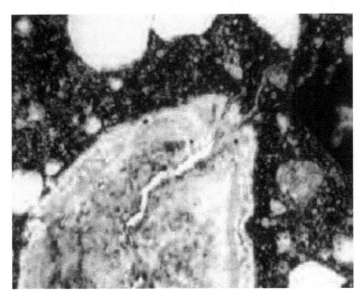

Figure 3.12 A crack in ASR-affected aggregate. (From Detwiler, R. J., and Taylor, P. C., "Specifier's Guide to Durable Concrete," Portland Cement Association, Skokie, IL, PCA EB 221, 2005. PCA image 13651. With permission.)

the environment. Lithium compounds have also been shown to be effective in preventing the distress because the gel formed is not expansive.

Any activity, including effective moist curing, that will help to reduce the ingress of water into the microstructure will also reduce rate of damage accumulation, but without SCMs or lithium are unlikely to be sufficient to prevent damage.

Physical mechanisms

Mechanical loading

Concrete is a brittle material that is strong in compression and weak in tension. Structural designers pay a lot of attention to ensuring that overload does not occur in a given element, and as such, this form of failure is rare unless imposed loads are far greater than expected or the quality of the mixture falls short of expectations.

Fatigue

Fatigue is the accumulation of damage over time in elements subjected to repeated stresses that are lower than the ultimate strength of the material. Concrete structures other than pavements are not normally considered from the point of view of fatigue, primarily because cyclic live loads are often small in comparison to the constant dead loads. However, the trend toward slender structures may force designers to be more aware of the potential for this type of premature failure.

Fatigue is a common problem in concrete pavements, particularly at the edges of panels where there is no adjacent panel to help redistribute loads. If there is a loss of support due to erosion or pumping, then the flexural loads can be high and the number of cycles to failure is dramatically reduced.

As the ratio of stress to strength decreases the number of cycles to failure increases, as shown in Figure 3.13. Stress is a function of the thickness of the slab, the stiffness of the support, and the magnitude of the load. The number of cycles depends on the required life of the system and the traffic (in particular the proportion of trucks). Designers can use these data and a model, such as the plot in Figure 3.13, to select slab thickness and required concrete strength. Refinements to the approach are to use approaches, such as Miners rule, to account for the variation in loads that will be applied to the system.

The effect of curing on fatigue resistance is limited to the impact on strength, particularly on the tensile strength at the top surface where cracks may initiate.

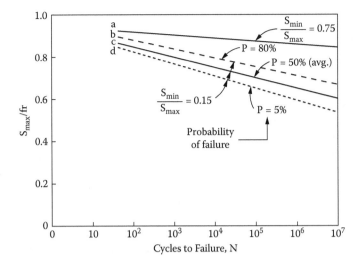

Figure 3.13 Typical S–N curve relating load–strength ratio to number of cycles to failure in a plain concrete beam. (From American Concrete Institute (ACI) Committee 215, "Considerations for Design of Concrete Structures Subjected to Fatigue Loading," 215R-74 (92) (Reapproved 1997), American Concrete Institute, Farmington Hills, MI, 2001. With permission.)

Abrasion, polishing, and erosion

Wear on concrete surfaces can occur on floors and slabs due to small-wheeled traffic, on pavements and slabs subject to vehicular traffic, and through erosion or cavitation on hydraulic structures.

Abrasion is initially resisted at the concrete surface; therefore, it is the hardness or strength of the paste at the surface that will govern the initial rate of wear until the aggregate is exposed. Most aggregates are harder than paste and fine aggregate hardness will start to govern the rate of wear once they are exposed. Paste strength may, however, still influence the rate of wear by controlling when aggregate particles are plucked out of the matrix.

Concrete that is likely to be subjected to abrasive loading should, therefore, be made with hard (fine) aggregates in a low water–cement ratio and well-cured concrete. The aggregate content should be reasonably high so that the thickness of the paste layer at the surface is kept to a minimum without compromising the surface finish. It is critical that the surface of the concrete be as strong as reasonable, which requires careful selection of finishing techniques for interior and exterior applications, and proper curing practice (Sawyer 1957). This need for surface strength must be balanced with the increasing risk of drying-related cracking associated with increasing strength.

Freeze–thaw and deicer scaling

Water expands approximately 9% when it freezes. Concrete that is saturated when the temperature drops below freezing is therefore subjected to expansive forces, leading to cracks parallel to the surface. Once a crack has opened, water will fill it when temperatures rise, and thus becomes available to jack the crack open farther on the next freezing cycle, eventually removing the surface of the concrete. The cycle then starts again below the new surface. This is why damage to concrete subjected to freezing and thawing is cumulative and can occur to some depth. The depth of damage for any given cycle is controlled by the saturation of the concrete and the depth to which freezing temperatures penetrate, meaning that cracking up to a hundred millimeters below the surface can occur in locations like water-retaining structures in locations with long, cold winters.

Concrete that is less than about 90% saturated is theoretically unlikely to be at risk of damage. However, saturation is often localized, meaning that measurement of overall moisture content is a poor predictor of performance. Moisture will tend to move toward zones that are already frozen and toward zones with higher salt concentrations (by osmosis), thereby setting high pressures and local variations in saturation on a mesoscale. Some deicing salts such as magnesium chloride will attract water rather than dry out, leading to greater saturation in the concrete (Figure 3.14).

The temperature at freezing is a function of the size of the void and the concentration of the salts in the pore solution. Smaller voids will tend to be filled with water at higher pressures due to surface tension effects; thus, the temperature at freezing is depressed. This means that water in larger voids

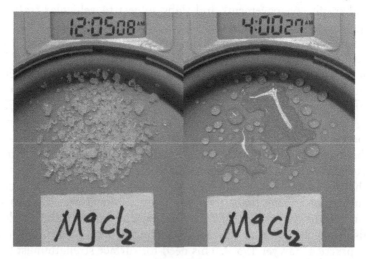

Figure 3.14 Two images of a sample of magnesium chloride salt that has dissolved in water absorbed from the atmosphere. (Courtesy of Jiake Zhang.)

will freeze at higher temperatures than water in smaller voids. Generally, water in capillary pores freezes at about −4°C (25°F), whereas water in the smaller gel pores freezes at about −78°C (−108°F).

Likewise, if salts are present in the pore water (for example, from deicing salts applied to pavements), the osmotic pressure is increased. The damage thus incurred is normally localized at the surface, resulting in a loss of surface concrete, referred to as scaling (Neville 1996).

The most effective methods of reducing the risk of damage in a freeze–thaw environment are as follows:

- Reduce the availability of water to saturate the concrete. This is not always possible, but it should be noted that concrete that is allowed to dry periodically is at a much lower risk (Figure 3.15). Good drainage detailing around the structure is essential to help prevent saturation of the concrete.
- Reduce the porosity of the system to minimize the freezable water available for expansion. This is achieved by using a sufficiently low water–cement ratio—less than 0.45—and adequate curing (ACI Committee 308 2001). Some authorities are specifying w/cm as low as 0.40 to reduce the risk of distress. Curing is essential to obtain the benefit of using low w/cm systems.

Figure 3.15 A slab at a car wash. The concrete in the driving lane around the building where wet cars leave the facility is severely distressed, whereas the same concrete slab not continually wet is in good condition.

- Provide sufficient space for the water to expand into in the form of a large number of small air bubbles. This is the motivation behind entraining air in concrete.

D-cracking

D-cracking is a similar phenomenon to that described in the previous section, except that the damage occurs within the coarse aggregate particles rather than in the paste. The problem occurs only in saturated porous aggregates. Research has shown that aggregates that have a particular pore size distribution are most at risk of distress. Large pores allow water to enter and leave the surface of the particles readily and are not a problem, while very small pores do not allow water to enter the system. However, pores in the range of 0.04 to 0.2 μm allow water to enter, but on freezing it does not escape, thus staying in the aggregates, freezing, expanding, and causing damage (Marks and Dubberke 1982).

The damage is most commonly observed in pavements, because water is drawn up to the underside of concrete slabs on the ground, saturating the concrete. As freezing and thawing cycles occur on the surface, more moisture is drawn up through the slab, particularly near sawn joints. Damage is often observed as D-shaped cracks near and starting at the bottom of joints (Figure 3.16). The cracks start in the aggregates and propagate into the paste.

Smaller particles expand less due to the smaller volume of pores (and water) within them, and therefore cause less damage. A common practice is to limit the maximum size of the coarse aggregate if the material is known or suspected to be prone to D-cracking, although this delays rather than prevents distress.

Prevention is only achieved by avoiding the use of at-risk aggregates although the more impermeable the system, the slower the rate of damage accumulation.

Reducing the risk

Exposure conditions

A fundamental part of understanding concrete deterioration is to consider the environment to which the concrete is exposed and to minimize its impact.

A critical parameter in concrete durability is the amount of moisture or water with which the concrete is in contact. All chemically based deterioration mechanisms require the presence of water to proceed; therefore, concrete in a dry environment is less likely to deteriorate than concrete in a wet environment. Often the water is a transport medium for aggressive chemicals in solution, such as sulfates or chlorides.

Some chemical deterioration or alteration mechanisms also require the presence of a gas such as oxygen (steel corrosion) or carbon dioxide

Figure 3.16 Typical D-cracking. (Courtesy of Jim Grove.)

(carbonation). Thus, a concrete that is continuously submerged may be at lower risk than concrete that is subject to cycles of wetting and drying.

It should be noted that there is no direct correlation between concrete strength and potential durability. Specifying a given compressive strength and hoping for a durable concrete may not have the desired result. For example, the mechanism of attack may be one that can be accommodated only by the use of significant amounts of supplementary cementing materials, which may result in slightly lower early strengths and higher later strengths.

Consideration must also be given to the local environment within the system. A concrete slab on grade may be subject to water being drawn to the underside through the ground if the air above is dry. This will result in a moisture gradient through the slab and may cause warping, as well as deposition of salts on or near the top surface as the moisture is carried through the concrete (and more commonly through the joints) by capillary action, and then evaporates. Another common attack mechanism is wicking, a process in which groundwater is carried up vertical elements and evaporates through surfaces exposed above ground. Wicking can result in very high concentrations of aggressive salts just above ground level in walls, causing chemical attack or salt scaling damage.

The "microclimate" of each concrete element should also be considered. A wall exposed to sea spray will have a much higher chloride concentration on the seaward side than on the leeward side. This will be a critical aspect when, for example, samples are taken from such structures. The temperature to which the concrete is exposed may also have a significant effect on durability. Structures such as bridge decks are more exposed than the approach pavement, meaning that the number of freeze–thaw cycles will be significantly higher. An environment that is cycled (wet–dry or hot–cold) is more severe than a constant environment. Increasing temperature will tend to increase the rate of chemical attack.

Transport properties

Because the presence of water or other fluids is critical to the aforementioned mechanisms, it is clear that a concrete system that can resist the passage of fluids will be in a much better position to survive the environment to which it is exposed. This section therefore discusses the transport properties of concrete.

Transport properties include both permeability (the ease with which fluids move through concrete) and diffusivity (the ease with which ions move through concrete). A concrete of low permeability is difficult to saturate with water, and thus is less likely to suffer significantly from frost damage. However, concrete of low permeability is not immune to chemical deterioration, but it suffers these effects only on the surface. Thus, a low permeability concrete lasts much longer than a highly permeable concrete in the same environment.

The time required for capillary pores to become discontinuous with increasing hydration decreases with reducing w/cm (Powers et al. 1959), and mixtures with a w/cm greater than 0.7 will always have continuous pores. This means that the higher the w/cm up to about 0.7, the longer the period of moist curing required for adequate permeability.

The use of supplementary cementing materials can significantly reduce the permeability and diffusivity of concrete. The greatest benefits from the standpoint of durability derive from the pozzolanic reaction—calcium hydroxide reacts with silica and water to form calcium silicate hydrate. Because calcium silicate hydrate has a greater volume than the calcium hydroxide and pozzolan from which it originates, the pozzolanic reaction results in a finer system of capillary pores.

As stated before, curing is the provision of sufficient water and appropriate temperatures for sufficient time for the cement hydration to proceed. It is critical for durability, in particular for near-surface concrete. A lack of curing will generally affect only the outer 30 to 50 mm (0.8 to 2 inches) of an element, but that critical zone is exposed to the environment and provides protection and passivation to the reinforcing steel. The concrete in this zone is sometimes referred to as "covercrete."

Concretes containing supplementary cementing materials are generally more sensitive to a lack of curing than those made with only portland cement. This is because the reaction of these materials is normally slower, although it will continue for longer. Concrete containing supplementary cementing materials and that is well cured will generally have lower permeability; if it is poorly cured, it will be more permeable than an equivalent portland-cement-only concrete.

Concretes with water–cement ratio above about 0.4 contain sufficient water to ensure full hydration of the cement (Addis and Alexander 1990). A lower water–cement ratio will result in all of the water being consumed before all of the cement is hydrated. Thus, it is sufficient to use systems that seal water into the concrete to cure higher water–cement ratio concretes. When dealing with lower water–cement ratio concretes, it may be preferable to provide additional water during the curing process.

STRENGTH

Fundamentals

Structural engineers are extremely concerned about the strength of a mixture. This is not inappropriate because strength is fundamental to structural integrity, but, as noted earlier, strength is not always directly related to durability, which in many ways is more critical to long-term serviceability of concrete structures.

Structural specifications assume that a given strength will be achieved as required, with little reference to how that strength may be achieved. Some specifications impose a belt and braces approach by limiting water cementitious materials ratios and cementitious materials contents in a mixture. Design codes make accommodation for variability and the risk of not achieving specified strength by statistical means, either by increasing the required strength based on standard deviations at the batch plant, or by applying multipliers or factors of safety to loads or strengths.

Strength is the ability of an element to resist load. If load exceeds strength, then failure is likely. Under loading, cracking will start at about 70% of the ultimate strength in conventional concrete, with stiffness decreasing as damage accumulates. For high strength mixtures, the damage becomes apparent at about 90% (Carrasquillo et al. 1981).

The nature of materials is such that strength in one direction may be different from strength in another, and concrete is a prime example. Plain concrete is strong in compression and weak in tension, therefore reinforcing steel is used in structural elements to carry tensile loads. Concrete is also brittle, meaning that it does not deform significantly after it starts to fail, but cracks form and grow rapidly. Like all performance properties, strength is a function of the mixture proportions, primarily the w/cm, and the degree of hydration of the cementitious materials.

Curing is often referred to in specifications because of concerns about strength, which may need to be considered with some caution as discussed later.

Mechanisms

Strength of concrete is provided by the hydration products of the cementitious system, that is, aggregates are glued together by hydrated cement paste. As discussed in Chapter 2, cement reacts with water to form a mixture of compounds. This reaction is initially very rapid, then slows but continues for a long time if there is unhydrated cementitious material and water in contact with each other. SCM reactions are slower but continue for longer, meaning that systems containing SCMs require more attention to curing to ensure adequate hydration.

The effects of drying, or cyclic wetting and drying, on strength are complex. On the one hand, premature drying reduces the degree of hydration of the system leading to reduced strength. On the other hand, once adequate hydration has been achieved, water remaining in the pores has a wedging effect thus reducing strength in small specimens (Popovics 1986). The other factor that influences performance is that internal stresses can be set up within the matrix as a result of variations in the moisture state across a section. These residual stresses can lead to decreased measured strength (Popovics 1986; Soroka and Baum 1994). This effect is more apparent in high-performance systems, likely due to lower permeability of the matrix leading to greater differentials across a section. The moisture state of test samples must therefore be rigorously controlled to reduce errors in standardized strength tests.

Curing and strength

It is not uncommon both in the literature (Alsayed 1994; Bushlaibi 2004) and in practice for engineers to be concerned about the effect of curing on strength. This is not surprising, because strength is considered the primary property of concrete that governs system performance, particularly in the context of structural design. It is also not surprising, because much of the data in the literature indicates that there is a marked drop in strength with reduced curing quality (Figure 3.17). In addition, strength reductions due to drying are reportedly greater for high-strength mixtures than normal-strength mixtures, particularly at early ages (Carrasquillo et al. 1981).

However, it has also been recognized since the 1980s (Meyer 1987) that the strength and curing relationship typically presented may not be entirely valid in structural-sized elements, because the relative volume of material

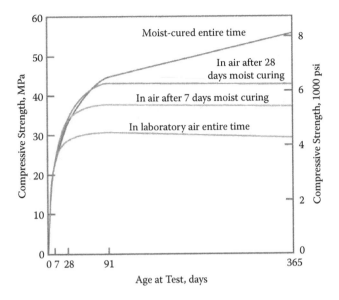

Figure 3.17 Typical data that relates compressive strength with curing time for small samples. (From Kosmatka, S. H., and Wilson, M. L., *Design and Control of Concrete Mixtures*, 15th ed., Portland Cement Association, Skokie, IL, 2011. With permission.)

affected by poor curing reduces with increasing specimen size. Most data available on the relationship between strength and curing is based on laboratory-sized specimens, typically 50 to 150 mm cubes.

Data by Soroka and Baum (1994) confirms that increasing specimen size shows a decreasing effect of curing (Figure 3.18). This is to be expected

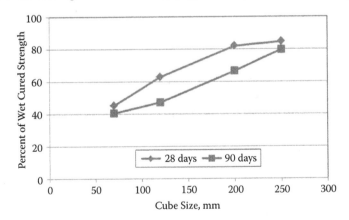

Figure 3.18 Increasing test sample size results in a reducing effect of poor curing in compression tests. (Data from Soroka, I., and Baum, H., *Journal of Materials in Civil Engineering*, 6, 1, 16–22, 1994.)

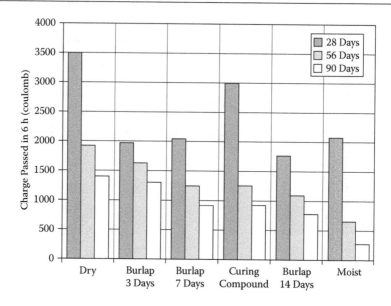

Figure 3.19 Data showing significant increase in chloride penetrability with poor curing. (From Nassif, H., and Suksawang, N., *Transportation Research Record* 1798-31, Paper No. 02-3305, 2002. With permission.)

because the quality of the curing will only affect the outer surface of the sample. As samples get bigger, the relative volume of material harmed by poor curing therefore gets smaller. Bentur and Goldman (1989) also noted that in 70 mm cubes, poor curing decreased strength by 20% but carbonation increased 100%. Similarly Nassif and Suksawang (2002) reported significant effects on durability yet smaller effects on strength (Figure 3.19). This is because strength is controlled by the quality of the total volume of the sample, whereas durability tests generally only consider the quality of the surface concrete.

All of this leads back to the point that curing is critically important but not for the reasons commonly accepted. As discussed in an earlier section, potential durability is strongly affected by poor curing in the field, even if in real structural-sized elements the effect on strength is relatively small.

Controlling strength

As noted earlier, the primary factor that controls strength of concrete is the w/cm of the mixture, cementitious chemistry, and the degree of hydration. The first two parameters are controlled at the time of material selection and proportioning. Final w/cm in the in-place concrete and hydration are construction-related issues, with curing playing a large part in influencing the hydration of the outer skin of the concrete.

In summary, curing may not influence structural strength significantly, whereas good design and construction practice including rigorous quality assurance will have a significant affect.

MODULUS OF ELASTICITY

Fundamentals

Modulus of elasticity is a measure of the stiffness of a material and is determined by measuring a change in stress (by changing applied load) divided by the associated change in strain (by measuring deformation).

Modulus of elasticity is important to design engineers for two reasons. First, deformations under applied loads will govern the serviceability of a structure. A road bridge that sags every time a vehicle goes over it may be strong enough, but users will be uncomfortable and safety may be compromised. Design of tall buildings may be controlled by the need to keep the natural frequency of the top floors out of the range detected by the human body leading to seasickness.

The second importance of stiffness is in the relationship between imposed strains due to temperature and moisture changes, and the associated risk of cracking, as discussed in the section on cracking. Systems with high stiffness are at greater risk than those that are more elastic. This is most marked in elements that are not reinforced such as slabs on grade.

Factors affecting stiffness

Similarly to the discussion on strength, the factors that control stiffness include the following:

- Water–cement ratio of the paste
- Degree of hydration of the paste
- Type of aggregate
- Relative proportions of paste and aggregate

Also, similarly to strength, the effect of curing is relatively small because stiffness is governed by the bulk properties of the element. Curing will generally only influence the surface zone, which may be relatively small in structural-sized elements.

CREEP

Fundamentals

Creep is defined by ACI as "a time-dependent increase in strain under constant load taking place after the initial strain at loading." It comprises two components (ACI Committee 209 2001):

- Drying creep—The creep occurring in a specimen exposed to the environment and allowed to dry
- Basic creep—Time-dependent increase in strain under sustained constant load of a concrete-sealed specimen

Drying creep is therefore the result of water movement through and out of the cementitious system. Basic creep is due to deformation of the cement matrix along with microcracking and damage accumulation.

In other words concrete under constant load will tend to deform. The consequence of this deformation is that stress is relieved over time. This is both a benefit in that stresses may be relieved as the concrete effectively moves away from the source of the load, but in prestressed systems the amount of prestressing force is reduced under creep, thus reducing the load-carrying capacity of the system. Increasing deflections over time can also be problematic in some structural systems, for instance, if a slender beam sags, then doors or windows under it will jam.

Factors affecting creep

Creep of concrete is affected by several variables:

- Moisture content has a strong influence because drying creep is due to movement of water in the paste. Increasing moisture content at the time of loading increases deformation, as there is more water available to be displaced.
- The stress–strength ratio at the time of loading is linearly related to the magnitude of creep.
- The water–cementitious materials ratio controls the strength, stiffness, and permeability of cement paste. Therefore, w/cm indirectly affects the creep of paste because changing w/cm leads to a change in strength and so the stress–strength ratio. A decrease in w/cm also reduces permeability and thus slows water movement.
- Cement type or blend affects creep largely due to differing rates of hydration and therefore strength and permeability.

- Aggregates reduce the creep of concrete by restraining paste movement. Concrete creep is therefore decreased with increasing aggregate volume and stiffness.
- Specimen geometry and size reflected by the volume–surface ratio affect the rate and magnitude of creep. Increasing values of this ratio will have slower moisture loss and consequently reduced drying creep.
- Drying conditions. Drying creep will increase with a decrease of ambient relative humidity or an increase in temperature due to more rapid water movement.

Controlling creep

The effect of curing on creep is therefore largely related to the degree of hydration of the system at the time of loading, and thus the permeability and strength. The other effect is on the moisture state at the time of loading. As noted in the discussion on strength, the effect of curing is likely limited because on larger structural elements the volume of concrete affected by poor curing is limited to the surface zone.

Creep can be reduced by loading when an appropriate strength has been achieved. Higher strengths should be specified if early loading is required.

SUSTAINABILITY

Definition

A basic definition of sustainability is the capacity to maintain a process or state of being into perpetuity. In the context of human activity, sustainability has been described as activity or development "that meets the needs of the present without compromising the ability of future generations to meet their own needs" (World Commission on Environment and Development [WCED] 1987). In other words, we seek to leave the planet for our children in the same or better condition than it is now.

Three general categories of needs and impacts are generally considered: environmental, economic, and social. Together, the three categories form a "triple bottom line" (Elkington 1994) of sustainability. No category should be under or overvalued, but rather a balance among the three must be found (Figure 3.20; Van Dam et al. 2011).

In any construction activity, the three considerations often compete. For example, society is dependent on the ability to freely move people, goods, and information. This is clearly illustrated when natural disasters break down the transportation infrastructure, and the quality of life in cities rapidly deteriorates. However, there is a large economic and environmental cost to providing and using that infrastructure.

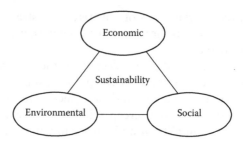

Figure 3.20 Representation of sustainability's "triple bottom line."

Improving sustainability in concrete systems is simply good engineering, which simply involves working with limited resources to achieve the best product possible. What has changed is the way the product is evaluated and the period of time over which it is evaluated. In the past, economic factors were paramount; now sustainability requires that environmental and social factors be considered over the full life of the system.

To balance these factors there is a need to measure and prioritize suitable parameters that provide realistic quantification of both impacts and benefits. Complicating this quantification process is that factors must be evaluated over the whole life of the constructed facility from conception to removal. Further, each category of sustainability is measured or rated in different terms: economic factors in terms of dollars; environmental factors in terms of global warming, or air and water quality; and social factors in terms yet to be determined. There is no obvious way to balance or weigh them.

It is important that improvements in sustainability continue to be developed and implemented. The challenge is to be imaginative, and consider what materials and techniques can be applied to further increase the economic, environmental, and social benefits in the concrete-based infrastructure (Van Dam et al. 2011).

Factors affecting sustainability

Different factors affect the three categories of the triple bottom line.

Economic

Engineers are used to balancing alternative approaches to design, materials, and construction to achieve the lowest initial cost. This is because most contractual systems are based on a low-bid approach.

To be truly sustainable it is important that economic assessments should be based on life-cycle cost analysis (LCCA) to capture the real costs over the lifetime of the structure. Models and methods are available to conduct this work (Santero et al. 2011). Some authorities are requiring that designs and systems be compared on this basis. One difficulty is that the validity of any calculation is extremely dependent on assumed discount rates, which are a topic of debate.

Environmental

A rigorous discussion about environmental effects in the built environment is a complex issue that is continually changing and well beyond the scope of this publication. Additional information can be obtained from Van Dam et al. (2011). However, some basic principles can be presented here.

A number of environmental parameters have to be considered beyond just the carbon dioxide burden. Environmental effects can be separated into

- The production of pollutants, including but not limited to
  - Greenhouse gases such as CO_2 and methane
  - Atmospheric contaminants such as NOx, volatile organic compounds (VOCs), and dust
  - Water contaminants such as heavy metals and organic compounds
- The consumption of nonrenewable materials including
  - High-quality aggregates
  - Oil-based products
  - Metal ores

In the context of concrete, most attention is focused on CO_2 equivalents because cement production is a significant contributor to worldwide anthropomorphic CO_2 generation. However, a large factor in this contribution is the sheer volume of cement consumed worldwide. Most of the CO_2 burden associated with a concrete mixture is from the cement, even though it only comprises about 10% of the mass of concrete. Therefore, any means of reducing cement content in a mixture, without compromising performance, will likely measurably reduce the impact of the structure being constructed.

There are three effective ways of reducing cement consumption in a structure (Mehta 1986):

- More efficient structural design leading to smaller elements
- Use of supplementary cementitious materials to reduce the amount of cement in a mixture
- Mixture proportioning that uses cementitious materials more efficiently

Of course, an obvious means of reducing environmental impact is to increase the longevity of a structure, thereby reducing the frequency with which it has to be repaired or replaced. Tied with this is the large benefit obtained by avoiding the impacts of delays and inconvenience to users associated with a breakdown in the transportation infrastructure. The use phase of built environment elements is often associated with greater impacts than the construction or removal phases (Wathne 2011). Compact communities that minimize transportation distances are therefore likely more sustainable than spread-out communities.

Decisions made at all stages of a structure's lifetime including design, materials selection, construction, use, remediation, and removal, will impact many of the aforementioned factors. It is also critical that when determining the impacts of different selections, the full life cycle should be accounted for.

Social

The social category is the hardest for engineers to quantify. Society is dependent on a working infrastructure system to deliver food; water; shelter; power; medical care; and access to work, home, and play locations. At the same time the system has to be able to remove and deal with waste products. When these systems break down, such as in natural disasters, the community quickly becomes desperate, often leading to rioting.

Points to consider are that there are real costs to society associated with

- Inadequate housing
- Traffic delays
- Inability to deliver critical needs efficiently

These factors can be measured in terms of direct financial costs as well as impacts on health and well-being.

Curing and sustainability

The impact of curing practices can be discussed in sustainability terms for all three of the critical categories.

The cost of curing is normally relatively low, whereas the benefits to durability are large. However, relatively few specifications and contracts effectively include curing as a pay item, because unless inspectors are on site continually, it is difficult to confirm after the fact whether curing has been conducted. As discussed in the next chapter, consideration may be given to calibrating the performance of a given mixture to the curing provided, and to use this factor as a pay item. If curing does have an effect on performance, it is likely to be on the potential durability of the mixture, and therefore the potential longevity of the structure. The cost implications of premature failure are huge.

The environmental benefits of curing are tied, again, to the longevity of the system. The impacts are dependent on the approach used to provide curing. If flooding is the method of choice, then questions have to be addressed about the disposal of the water afterward. If curing compounds are used, as is most common, then the type and impact of the transportation medium, whether water or volatile-organically based, have to be included in assessment of the whole system. The amount of materials used is usually relatively small; therefore the impacts should also be low.

Social impacts include, once more, the potential longevity of the structure as well as health effects on workers applying compounds to the concrete surfaces.

Overall it would appear that although some negative impacts are possible, they are likely relatively small because of the limited amount of chemicals used, whereas the benefits or increased longevity and more efficient usage of cements are likely to be large.

REFERENCES

American Concrete Institute (ACI) Committee 207, 1996, "Mass Concrete," ACI 201.1R-96, American Concrete Institute, Farmington Hills, MI.

American Concrete Institute (ACI) Committee 209, 2001, "Report on Factors Affecting Shrinkage and Creep of Hardened Concrete," 209.1R-05, American Concrete Institute, Farmington Hills, MI.

American Concrete Institute (ACI) Committee 215, 2001, "Considerations for Design of Concrete Structures Subjected to Fatigue Loading," 215R-74 (92) (Reapproved 1997), American Concrete Institute, Farmington Hills, MI.

American Concrete Institute (ACI) Committee 221, 2001, "A Guide for Use of Normal Weight and Heavy Weight Aggregates in Concrete," ACI 221R-01, American Concrete Institute, Farmington Hills, MI.

American Concrete Institute (ACI) Committee 224, 1993, "Causes, Evaluation and Repair of Cracks in Concrete Structures," ACI 224.1R-93, American Concrete Institute, Farmington Hills, MI.

American Concrete Institute (ACI) Committee 308, 2001, "Guide to Curing Concrete," ACI 308R-01, American Concrete Institute, Farmington Hills, MI.

Addis, B. J., and Alexander, M. G., 1990, "A Method for Proportioning Trial Mixes for High Performance Concrete," High Strength Concrete, SP 121, American Concrete Institute, Farmington Hills, MI.

Alsayed, S. H., 1994, "Effect of Curing Conditions on Strength, Porosity, Absorptivity, and Shrinkage of Concrete in Hot and Dry Climate," *Cement and Concrete Research*, vol. 24, no. 7, pp. 1390–1398.

Altoubat, S. A., and Lange, D. A., 2001, "Creep, Shrinkage, and Cracking of Restrained Concrete at Early Age," *ACI Materials Journal*, vol. 98, no. 4, pp. 323–331.

Ballim, Y., Taylor, P. C., and Lampacher, B. J., 1994, "Assessment and Control of Concrete Durability," Concrete Meets the Challenge, National Convention of the Concrete Society of Southern Africa, September.

Bentur, A., and Goldman, A., 1989, "Curing Effects, Strength and Physical Properties of High Strength Silica Fume Concretes," *Journal of Materials in Civil Engineering*, vol. 1, no. 1, pp. 46–58.

Bushlaibi, A. H., 2004, "Effects of Environment and Curing Methods on the Compressive Strength of Silica Fume High-Strength Concrete," *Advances in Cement Research*, vol. 16, no. 1, pp. 17–22.

Carrasquillo, R. L., Slate, F. O., and Nilson, A. H., 1981, "Microcracking and Behavior of High Strength Concrete Subject to Short-Term Loading," *Journal of the American Concrete Institute*, vol. 78, no. 3, pp. 179–186.

Chatterji, S., 1982, "Probable Mechanisms of Crack Formation at Early Ages of Concretes," International Conference on Concrete at Early Ages, RILEM, pp. 35–38.

Detwiler, R. J., and Taylor, P. C., 2005, "Specifier's Guide to Durable Concrete," PCA EB 221, Portland Cement Association, Skokie, IL.

Elkington, J. 1994. "Towards the Sustainable Corporation: Win-Win-Win Business Strategies for Sustainable Development," *California Management Review*, vol. 36, no. 2, pp. 90–101.

Famy, C., and Taylor, H. F. W., 2001, "Ettringite in Hydration of Portland Cement Concrete and Its Occurrence in Mature Concretes," *ACI Materials Journal*, vol. 98, no. 4, pp. 350–356.

Gollop, R. S., and Taylor, H. F. W., 1992, "Microstructural and Microanalytical Studies of Sulfate Attack. I. Ordinary Portland Cement Paste," *Cement and Concrete Research*, vol. 22, no. 6, pp. 1027–1038.

Haynes, H., O'Neill, R., and Mehta, P. K., 1996, "Concrete Deterioration from Physical Attack by Salts," *Concrete International*, vol. 18, no. 1, pp. 63–68.

Holt, E. E., 2000, "Where Did These Cracks Come From?" *Concrete International*, vol. 22, no. 9, pp. 57–60.

Holt, E. E. 2001, *Early Age Autogenous Shrinkage of Concrete*, Technical Research Centre of Finland, VTT Publications 446, Espoo, Finland.

Kosmatka, S. H., and Wilson, M. L., 2011, *Design and Control of Concrete Mixtures*, 15th ed., Portland Cement Association, Skokie, IL.

Marais, L. R., and Perrie, B. D., 1993, *Concrete Industrial Floors on the Ground*, Cement and Concrete Institute, Halfway House.

Marks, V. J., and Dubberke, W., 1982, "Durability of Concrete and the Iowa Pore Test," Transportation Research Record 853, Washington, D.C., pp. 25–30.

Mehta, P. K., 1986, *Concrete: Structure, Properties, and Materials*, Prentice-Hall, Englewood Cliffs, NJ.

Meyer, A., 1987, "The Importance of Surface Layer for the Durability of Concrete Structures," ACI Special Publication 100, John M. Scanlon, ed., vol. 2, American Concrete Institute, Detroit, MI, pp. 49–61.

Miller, F. M., and Conway, T., 2000, "Use of Ground Granulated Blast Furnace Slag for Reduction of Expansion Due to Delayed Ettringite Formation," ASTM C01/C09 Symposium on Prescriptive and Performance Specifications for Hydraulic Cements and Their Use in Concrete: Issues and Implications for Standards Development, American Society for Testing and Materials, West Conshohocken, PA, December 6.

Mindess, S., and Young, J. F., 1981, *Concrete*, Prentice-Hall, Englewood Cliffs, NJ.

Nabil, B., Aissa, A., and Aguida, B., 2011, "Use of a New Approach (Design of Experiments Method) to Study Different Procedures to Avoid Plastic Shrinkage Cracking of Concrete in Hot Climates," *Journal of Advanced Concrete Technology*, vol. 9, no. 2, pp. 149–157.

Nassif, H., and Suksawang, N., 2002, "Effect of Curing Methods on Durability of High-Performance Concrete," Transportation Research Record 1798-31, Paper No. 02-3305.

National Ready Mixed Concrete Association (NRMCA), 1992, "What, Why & How? Plastic Shrinkage Cracking," CIP 5, National Ready Mixed Concrete Association, Silver Spring, MD.

Neville, A. M., 1996, *Properties of Concrete*, 4th ed., John Wiley & Sons, New York, NY.

Popovics, S., 1986, "Effect of Curing Method and Final Moisture Condition on Compressive Strength of Concrete," *Journal of the American Concrete Institute*, vol. 83, no. 4, pp. 650–657.

Poole, T. S., 2006, "Curing Portland Cement Concrete Pavements, Volume II," FHWA-HRT-05-038, Federal Highway Administration, McLean, VA.

Powers, T. C., Copeland, H. E., and Mann, H. M., 1959, "Capillary Continuity or Discontinuity in Cement Pastes," Research Department Bulletin RX110, reprinted from the *Journal of the PCA Research and Development Laboratories*, vol. 1, no. 2, pp. 38–48.

Rostásy, F. S., Tanabe, T., and Laube, M., 1998, "Assessment of External Restraint," in *Prevention of Thermal Cracking in Concrete at Early Ages*, RILEM Report 15, R. Springenschmid, ed., E. & F.N. Spon, London, pp. 149–177.

Santero, N. J., Masanet, E., and Horvath, A., 2011, "Life-Cycle Assessment of Pavements. Part 1: Critical Review," *Resources, Conservation, and Recycling*, vol. 55, no. 9–10, pp. 801–809.

Sawyer, J. L., 1957, "Wear Test on Concrete Using the German Standard Method of Test Machine," *Proceedings of ASTM*, vol. 57, pp. 1145–1153.

Shimada, Y. E., 2005, "Chemical Path of Ettringite Formation in Heat Cured Mortar and Its Relationship to Expansion," Ph.D. thesis, Northwestern University, Evanston, IL.

Soroka, I., and Baum, H., 1994, "Influence of Specimen Size on Effect of Curing Regime on Concrete Compressive Strength," *Journal of Materials in Civil Engineering*, vol. 6, no. 1, pp. 16–22.

Taylor, P. C., Johansen, V. C., Graf, L. A., Kozikowski, R. L., Zemajtis, J. Z., and Ferraris C. F., 2006, "Identifying Incompatible Combinations of Concrete Materials, Volume I–Final Report," Report HRT-06-079, Federal Highway Administration, Washington, D.C.

Taylor, P. C., Tikalsky, P., Wang, K., Fick, G., and Wang, X., 2012, "Development of Performance Properties of Ternary Mixtures: Field Demonstrations and Project Summary," DTFH61-06-H-00011 Work Plans 12 and 19, Federal Highway Administration.

Tongaroonsri, S., and Tangtermsirikul, S., 2009, "Effect of Mineral Admixtures and Curing Periods on Shrinkage and Cracking Age under Restrained Condition," *Construction and Building Materials*, vol. 23, pp. 1050–1056.

Tracy, S. L., Boyd, S. R., and Connolly, J. D., 2004, "Effect of Curing Temperature and Cement Chemistry on the Potential for Concrete Expansion Due to DEF," *PCI Journal*, pp. 46–57.

Van Dam, T., Taylor, P. C., Fick, G., Gress, D., VanGeem, M., and Lorenz, E., 2011, "Sustainable Concrete Pavements: A Manual of Practice," National Concrete Pavement Technology Center, Ames, IA.

Wathne, L., 2010, "Sustainability Opportunities with Pavements: Are We Focusing on the Right Stuff?" 11th International Symposium on Concrete Roads 2010, Seville, Spain, October 13–15.

World Commission on Environment and Development (WCED), 1987, "Our Common Future: The Report of the World Commission on Environment and Development," Oxford University Press, New York, NY.

Chapter 4

Curing in practice

INTRODUCTION

As discussed in the earlier chapters, the primary purpose of curing concrete is to provide an environment that will support hydration of the cementitious materials. By doing this, the most value is obtained from the mixture and the performance of the system can be enhanced. The desired environment is that the mixture is wet enough and warm enough, but not too hot and without excessive variations. This chapter discusses practical approaches to providing this preferred environment as well as some discussion on how long is long enough.

SELECTING CURING METHODS

Three basic approaches to moisture control may be considered: water added to the surface (after final set), prevention of moisture loss, or providing water internally. Selecting a curing method will depend on a number of factors, including the geometry of the structure, the need for construction access during curing, the weather, and the mixture.

Accounting for structure type

Different types of elements may require different forms of curing.

Vertical surfaces

In general, vertical surfaces are less prone to environmental attack, because, unless they are immersed in water or in a tidal zone, they are rarely saturated. On the other hand, depending on the type of structure, there may be tight aesthetic demands on the finish or cladding systems that will be applied. In addition, the need to continue building the next lift or element may also limit the options available.

The basic rules are as follows:

- Keep forms in place as long as required for structural purposes and longer if possible for curing purposes
- Apply curing compound or protective layers as soon as forms are removed
- Avoid thermal shock

Structural elements like columns and walls may be more efficiently protected by leaving their forms in place, or by wrapping them in plastic and burlap. Such elements may also have architectural requirements that will govern the selection of pigmented materials.

Horizontal surfaces

Horizontal surfaces, particularly for slabs and slabs on grade, are generally particularly challenging, largely because the surface-to-volume ratio is high, surface areas are large, and construction traffic needs to get on the surface early.

The basic rules are as follows:

- Protect the surface from evaporation between leveling and final finishing using evaporation retarders, windbreaks, and fog sprays
- Do not finish the surface until bleeding is complete
- Apply curing compound of protective surfaces as soon as finishing is complete
- Protect from temperature extremes and rain

Application of curing compounds or sealants before bleeding has ceased may result in lenses of water being trapped below the surface. These lenses subsequently form layers of weakness leading to later cracking (Figure 4.1).

Figure 4.1 Spalling caused by weakened layers formed by bleed water trapped under a curing compound applied too early. (From Poole, T. S., "Curing Portland Cement Concrete Pavements, Volume II," FHWA-HRT-05-038, Federal Highway Administration, McLean, VA, 2006.)

Concrete floors with coverings

A special case of ensuring that concrete is in an appropriate moisture state is in new concrete floors that will have a covering bonded to them using adhesives. On the one hand the mixture must be kept moist for long enough to ensure that the required properties of strength and stiffness are achieved. On the other hand, the floor must dry sufficiently, because most adhesives are water based and will decompose if moisture is coming up through or from the concrete. In addition, if salts are transported to the surface they can deposit under the covering and debond it. Water trapped under an impermeable covering will also promote microbial growth leading to odors and potential health hazards.

This mixed requirement sets up a problem for the construction team. The following are recommended to balance the needs of the concrete and the covering (Farny and Tarr 2008; Kanare 2008):

- Ensure that there is adequate drainage beneath the slab to minimize the amount of groundwater under the slab.
- Use a low water–cement ratio (w/cm) mixture (<0.4) that will use most or all of the mix water in the hydration process. Such a mixture will also limit the amount of water that can rise under capillary forces through the slab.
- Wet cure or use curing compounds until the required properties of the concrete are achieved, then remove curing compounds if they have been used.
- Allow the slab to dry using heat and dehumidifiers as necessary. Do not use unvented heaters that raise the CO_2 level in the room.
- Test the moisture state of the surface before applying coatings. Test methods include a polyethylene sheet test (ASTM D4263-83), electrical resistance meter, nuclear moisture gauge, calcium chloride test (ASTM F1869-11), or in situ relative humidity (RH) meters (ASTM F2170-11).

Pavements

Pavements are sensitive to the effects of curing because of their large unformed surface areas, and because they are not always reinforced. Slipform pavement mixtures also tend to be very low slump and do not bleed significantly; therefore protection often needs to be applied shortly after it is exposed. However, the cure sprays often being mounted on the same bridge as the texturing, which has to wait for some stiffening of the mixture before it passes over the surface. There is also often a need to get construction traffic onto pavements within a day or two of placing, meaning that the curing system has to be resilient to heavy vehicle traffic. Curing compounds appear to be the most practical means of balancing the needs at minimum cost for such systems.

It is not common to pay particular attention to curing the saw cut faces of joints in slabs on grade. However, increasing use of aggressive deicing salts in cold climates is leading to increased rates of deterioration of joints. It is therefore recommended that as soon as the joint has been sawn, a curing compound should be applied to the fresh face. The down side of this is that sealants or fillers to be placed in the joint at a later stage will not bond to the surface. The design engineer will have to make a decision regarding detailing of these systems.

Pervious concrete

Pervious concrete comprises a mixture with little or no fine aggregate. The purpose of using such mixtures is to allow water to drain through the slab, thus reducing stormwater runoff loading and helping to replenish groundwater supplies. Because of the open nature of the system, premature drying is a high probability. Care must therefore be taken to ensure that the full depth of the slab is kept moist to allow hydration to proceed. Because of its design, ponding is not feasible. Therefore, pervious concrete should be misted with water then covered with polyethylene film secured at all the edges for several days. Superabsorbent polymers used as internal curing reservoirs have also been shown to be effective (Kevern and Farney 2012).

Accounting for concrete materials and mixture proportions

The cementitious system will affect some of the critical factors pertaining to curing needs, including strength gain rate, bleeding rate, and the period over which bleeding occurs.

Different cement grades and blends will hydrate at different rates. Finer cements will tend to hydrate faster, while systems containing supplementary cementitious materials will tend to be slower. The faster setting occurs, the sooner bleeding ends, thus accelerating the need to apply curing measures. It also means that if strength is being used as the criterion to end curing, then faster systems, such as rapid set cements, can be stopped sooner. This is beneficial if the curing method impedes later construction activities.

On the other hand, systems containing supplementary cementitious materials, particularly at higher dosages will generally tend to set later and take longer to gain strength. The risk of plastic shrinkage therefore increases, because curing compounds cannot be applied until after bleeding has ended but evaporative drying may be continuing. Likewise, strength gain is normally slower than in mixtures containing plain portland cement, meaning that curing is needed for a longer period. Fly ashes reportedly reduce the rate of bleeding, although it may continue for longer, and silica fume mixtures typically exhibit no bleeding, thus significantly increasing the need for protection from plastic shrinkage cracking in slabs on grade or at the tops of forms.

Admixtures used to adjust workability of a mixture may also affect setting time, depending on the chemistry of the system. Lignin or sugar based products may retard the system, particularly if overdosed, but polycarboxylate based products are likely to have less of an effect on setting. Set modifying admixtures will act as intended, either retarding or accelerating depending on the type of material used and dosage.

If the water–cement ratio is greater than about 0.40, then there is sufficient water in the original mixture to hydrate all the cement; therefore no added water is required. However, concretes with water–cement ratios less than 0.40 tend to require added water, because there is insufficient water in the mixture to hydrate all the cement. This can prove difficult, however, because the pores will be discontinuous after about 3 days, thus limiting the ability to transport the added water deep into the system. Added water will, however, significantly enhance the durability of the surface layer.

Although bleeding is primarily controlled by the gradation of the fine aggregate, low w/cm mixtures and low binder content mixtures will tend to exhibit less bleeding.

Accounting for the weather

The curing method should be balanced with the weather at the time the concrete is being cured, remembering that the aim is to keep the concrete warm and moist for the whole period.

Factors to consider include the following:

- Curing compounds will not provide any protection for temperature swings.
- Added water techniques have to be continuous, particularly in hot or windy conditions, or they should not be used.

Actions taken in hot weather should be focused on cooling the concrete and protecting it from rapid evaporation. The risk of cracking of slabs on grade is high in hot weather because of drying effects and stresses set up by temperature and moisture differentials. Software packages (McCullough and Rasmussen 1999) are available that help predict the risk of cracking based on modeling distribution of temperatures and moisture states in the slab from inputs of concrete mixture details and predicted weather.

Added water approaches may be acceptable in hot weather, but care must be taken to prevent thermal shock. The water should not be more than 10°C to 15°C cooler than the concrete (ACI Committee 305 1999). Larger elements that may get hot from internal heat of hydration should be protected from differentials greater than 10°C between the interior and the surface, and if the chemistry of the cementitious system is conducive to delayed ettringite formation, the system should be kept below 70°F.

Actions taken in cold weather should be focused on warming the concrete to enhance hydration. Risks for structural elements are that forms are released too early, resulting in failure or thermal shock. Risks for slabs on grade are related to cracking because of insufficient hydration to resist drying stresses, and freezing or freeze–thaw related distress.

Added water methods are not recommended in cold weather because of the risk of ice forming on the surface.

A single freezing event should be avoided until concrete has reached a strength of 3.5 MPa (ACI Committee 306 1988; ACI Committee 308 2001). Freezing and thawing cycles should be avoided until strength reaches 10 to 25 MPa (American Association of State Highway and Transportation Officials [AASHTO] 1998).

TEMPERATURE CONTROL

Curing is more than just keeping concrete moist but also entails controlling the temperature, both when the environment is too hot and too cold. It should be noted that the microstructure of slower (cooler) hydrating systems is normally finer, thereby leading to better permeability in the long run (Kjellsen et al. 1991).

Heating

Concrete placed in cold weather may need some heating or insulating to encourage hydration and to protect it from freezing.

As discussed in Chapter 2, hydration is an exothermic reaction. In addition, hydration is accelerated with increasing temperature, roughly doubling in rate with each 10°C rise. Concrete can therefore be placed at low temperatures, and if insulated, may still perform satisfactorily. The lower the temperature the slower the reaction, meaning that forms may have to stay in place for longer. Hydration proceeds at a much slower rate when the concrete temperature is below 4°C (40°F). Temperatures at or below −10°C in the pore system limit hydration such that little or no strength develops.

At low temperatures, slabs on grade should be protected for longer from wind and drying to prevent plastic shrinkage cracking (Figure 4.2). Such protection may be in the form of evaporation retarders, windbreaks, and fog sprays (discussed later). Below about −4°C there is a risk of freezing, resulting in the formation of ice crystals in the pores that leave characteristic patterns (Figure 4.3). Strength and durability in this case may be impaired and may not be recovered.

Figure 4.2 Typical plastic shrinkage cracks.

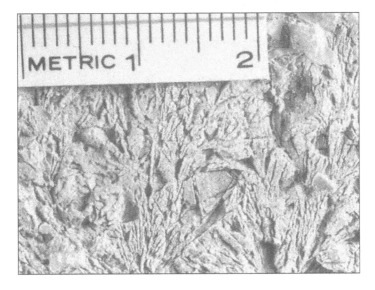

Figure 4.3 Markings left in concrete that froze when it was fresh. (From Kosmatka, S. H., and Wilson, M. L., *Design and Control of Concrete Mixtures*, 15th ed., Portland Cement Association, Skokie, IL, 2011. With permission.)

Protection in cold weather can either be passive or active. Passive protection is normally provided in the form of blankets that act as insulation layers keeping hydration heat in the system. These are particularly important for slabs on grade that have a large surface area and can be exposed to rapidly changing weather. It is not uncommon for an evening cold front to cause random cracking in floor slabs or pavements placed during a hot day. A mixture placed in the morning or early afternoon is likely to be at its hottest due to heat of hydration in the early evening, just when ambient temperatures may fall rapidly. The drop in temperature causes a large movement and high stresses in the immature and still weak system, leading to a high risk of cracking. Blankets should be made available when the weather report predicts a cold front and placed on the concrete well before the temperature starts to drop.

Dry, porous materials such as straw, burlap, or geotextile can be used as blankets. There are no specifications for blankets but they typically are in the form of multiple layers of tarpaulin with an insulating layer between them (Figure 4.4). Suitable insulating blankets are manufactured of fiberglass, sponge rubber, cellulose fibers, mineral wool, vinyl foam, and open-cell polyurethane foam. When insulated formwork is used, care should be taken to ensure that concrete temperatures do not become excessive. Curing concrete in cold weather should follow the recommendations in ACI 306 (ACI Committee 306 1988).

Blankets may also be applied to formed surfaces, but they are more often used to reduce the temperature gradient in mass concrete elements, and this is discussed in the next section.

Figure 4.4 Multilayer blankets. (Portland Cement Association image 165-15. With permission.)

Figure 4.5 A heater that is vented to prevent CO_2 buildup over the young concrete. (Portland Cement Association image. With permission.)

Active heating comprises using heaters to warm the early hydrating mixture (Figure 4.5). Again this is most common for slabs on grade where the timing of finishing activities is delicate. Care must be taken to avoid using unvented heaters burning oil-based compounds, because these devices emit large amounts of CO_2. Elevated CO_2 in a closed environment is both a risk for workers and can cause rapid carbonation of the young concrete surface. At an extreme this will lead to the surface being weak and appearing dusty because it lacks resistance to abrasion (Figure 4.6).

Figure 4.6 Dusting as the result of excessive carbonation and poor surface hydration. (Portland Cement Association image 1297. With permission.)

Another form of active temperature control is the use of steam to accelerate hydration. This practice is common in the precast industry where it is desirable to minimize the amount of time that the mixture remains in the forms. Care should be taken to avoid applying steam too early or too fast. Early and rapid heating will set up high gradients and so stresses in the elements leading to cracking. Likewise when steaming is completed the rate of cooling should be controlled. Guidance is provided in Figure 4.7 (Kosmatka and Wilson 2011).

Effective steam curing can lead to high early strengths. However, systems that have been steam cured typically do not continue to develop much additional strength over time; therefore factors of safety may have to be adjusted. In addition, permeability of steam-cured elements is reportedly higher, meaning that mixture proportions may have to be adjusted to achieve the desired durability. Mixtures containing fly ash and slag cement respond well to steam curing in that the benefit gained is higher than that obtained in a plain cement mixture.

Steam curing systems range from tightly controlled systems in sophisticated plants to a hosepipe running under a tarpaulin from a large kettle on a fire.

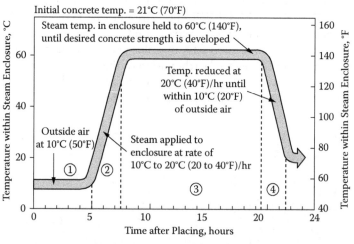

① Initial delay prior to steaming 3 to 5 hours
② Temperature increase period 2½ hours
③ Constant temperature period 6 to 12 hours*
④ Temperature decrease period 2 hours
  *Type III or high-early-strength cement,
  longer for other types

Figure 4.7 A recommended steam curing cycle. (From Kosmatka, S. H., and Wilson, M. L., *Design and Control of Concrete Mixtures*, 15th ed., Portland Cement Association, Skokie, IL, 2011. With permission.)

A topic that has been under some discussion among experts is that of delayed ettringite formation (DEF). General understanding of the issue is that mixtures with a particular chemistry and that have been heated above a critical temperature in the first few hours are prone to later age expansion and cracking. The critical temperature varies, but is typically about 70°C to 75°C. It should be noted that in some cases, large concrete elements might reach these temperatures without steaming but just by internal heat of hydration. The mechanisms of the distress are not fully understood but involve the dissolution of ettringite that is normally formed during early hydration. Months or years later, and in the presence of abundant water, the ettringite recrystallizes in voids and capillaries causing expansion and cracking. Preventive actions include maintaining careful control of the maximum temperatures experienced by the mixture and avoiding cementitious systems with a SO_3-to-Al_2O_3 ratio >0.5 (Mindess et al. 2003). Slag cement has been demonstrated to be effective at preventing the reaction (Miller and Conway 2003).

Cooling

As noted earlier, concrete that is placed at elevated temperatures may also need some close attention to prevent problems. The most significant issues are those related to uncontrolled setting, cracking, and a potential drop in long-term strength and durability.

As discussed in Chapter 2, incompatibility is a term applied to mixtures in which uncontrolled stiffening and setting may occur. Typically the phenomenon is observed when there is a combination of elevated temperatures, cements that may be marginally undersulfated, fly ashes containing high CaO contents, and use of lignin or sugar-based water-reducing admixtures. The reactions are complex, but in general terms the high temperatures and admixtures accelerate so-called flash-set reactions of the aluminates in the cement and fly ash. Normally these reactions are controlled by sulfates added to the cements, but the combinations of materials and conditions may let the system get out of control, leading to a spike in temperature and permanent stiffening of the mixture. Changing any one of the contributing factors will often prevent recurrence of the problem. One of these changes may include reducing the temperature of the mixture as discussed later.

The other major concern with elevated temperatures is the associated increase in risk of cracking. Large elements, typically those larger than about 1 m in minimum dimension, reach high temperatures at their cores because the heat cannot escape from the surface. This can then set up high differentials between the interior hot portion and the cooler surface that is at the same temperature as the environment. These differentials set up stresses leading to cracking, typically when the difference in temperature is greater than about 20°C (Poole 2006).

Data from the Texas Rigid Pavement (TRP) database for continuously reinforced pavement was analyzed by Schindler and McCollough (2002) to evaluate the effect of concrete temperatures on long-term performance. It revealed that there is an increased risk of failures with increasing air temperature during placement, especially above 90°F.

Activities to reduce the peak temperature may include the following:

- Chilling the concrete at the time of mixing by shading solid ingredients and their storage
- Sprinkling aggregate stockpiles
- Using chilled water or ice in the mixture
- Adding liquid nitrogen to the mix (Figure 4.8)

Specifications that limit maximum placement temperatures should account for the materials in the mixture and how they influence the following priorities that are strongly influenced by temperature including rate of hydration, thermal stresses, and permeability.

The lower the temperature of the mixture at the time of mixing, the lower the peak temperature. Use of relatively high dosages of appropriate supplementary cementitious materials will also help to reduce peak temperatures. If practical, it may be preferable to place the concrete at night when ambient

Figure 4.8 Liquid nitrogen being used to cool a batch of concrete. (Portland Cement Association image 12357. With permission.)

temperatures are cooler. Also as noted earlier, concrete placed late in the morning is likely to reach its peak temperature late in the afternoon when ambient temperatures are also at their highest, thus compounding the effect.

In mass concrete elements such as dam walls, temperatures may be reduced by embedding water pipes in the structure and running cold water through them. Another practice is to place blankets around the forms to insulate the system from the surroundings. The mixture still reaches relatively high temperatures, but the differential between the core and the surface is reduced, thus minimizing cracking. Cooling of very large elements may take from days to weeks. Extensive guidance is given in ACI 207 (ACI Committee 207 1996). Care should be taken when removing insulators because a shock temperature differential may be induced leading to cracking if the ambient temperature is 20°F lower than the concrete.

Another temperature-related problem is sometimes seen when a wall is formed above an older concrete foundation or wall without sufficient joints in it. Differential stresses caused by free movement at the top and restraint at the bottom can result in regular vertical cracks starting at the bottom but not reaching the top. This can be seen in slipformed elements such as barriers in highways. Risk can be reduced by taking efforts to reduce differentials in moisture and temperature state through the height of the wall, or by providing joints at regular intervals.

The structural design of slabs on grade is also strongly influenced by temperature effects. Concrete sets, that is, changes from liquid to solid, at about the same time the mixture reaches its maximum temperature. It therefore reaches its maximum dimension while still comprising a weak solid. Subsequent cooling, often through one face, can lead to very high stresses. The differential may also cause the slab to curl and lift off the substrate, thus adding self-weight to the stress pattern along with any added load due to vehicles.

It is therefore desirable to minimize the peak temperature of the mixture and to protect it from rapid cooling such as a cold front. Temperature effects are normally critical for the first 24 hours of the life of a slab. Subsequent diurnal and seasonal cycling may continue to exercise the slab, which combined with drying shrinkage will also add to differential stresses in the system. It is becoming apparent that the stiffer the foundation layer under a slab on grade, the greater the risk of temperature-related cracking, because as the slabs curl, the support becomes a point load rather than distributed loading.

Guidance on placing and caring for concrete in hot weather can be found in ACI 305 (ACI Committee 305 1999). Computer software packages such as HIPERPAV are available to assist the engineer in assessing the risk of problems based on the weather, mixture, and slab details. The package plots the stresses predicted in a system against the strength of the mixture, and flags when cracking is likely to occur (Figure 4.9).

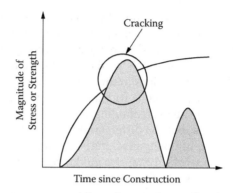

Figure 4.9 Typical output from HIPERPAV showing when cracking may occur. (From FHWA.)

Elevated temperatures will tend to accelerate hydration, and thus strength gain, initially, but performance will subsequently tend to plateau with little added gains over time. In addition, elevated temperatures have been shown to increase permeability. Mixtures developed for very high performance will often contain retarding admixtures to slow hydration and thus achieve a finer grain microstructure and better long-term properties.

MOISTURE CONTROL

Control of the moisture state of young concrete can have an enormous influence on its long-term performance. As discussed before, the properties of a mixture are governed by the degree of hydration of the cementitious system. Hydration, as the name implies, is a chemical reaction between water and the cementitious materials. If there is insufficient water in the mixture, then not all of the cementitious materials will react, leaving the mixture performing below its potential. It is therefore critical that sufficient water be made available to every cement grain for several days in order to get the most out of the mixture. Over time, more cement is hydrated and the rates of reaction slow significantly, and the benefits of continued curing decrease.

Work conducted by Senbetta and Malchow (1987) investigated the effects of curing methods on durability-related properties. The properties measured included abrasion resistance, scaling resistance, corrosion resistance of steel, chloride ion penetration, shrinkage, and absorptivity. The data showed that the most effective method, in terms of concrete performance, was sealing, followed by wrapping in plastic sheet, with the use of fog and the curing compounds showing similar performance. Other conclusions included that curing practices can significantly affect abrasion resistance, drying shrinkage, and capillary porosity.

The curing requirements for high-strength concrete may not be the same as those for more ordinary mixtures. Carrasquillo and Carrasquillo (1988) investigated this question looking at a wide range of mixtures subjected to three different curing conditions: fog room, curing compound, and none (air dry). Compressive strengths of cylinders showed improved strengths with moist curing. In addition, flexural strengths were more sensitive to curing than compressive strength. The difference was attributed to the effect of curing being a largely surface effect, and flexural strength is more significantly affected by surface condition of the sample.

If moist curing is interrupted, hydration continues until the concrete's internal relative humidity drops to about 80%. If moist curing is resumed, then hydration may resume, but it is difficult to resaturate concrete.

Water loss may occur in two ways. If the surface is exposed to drying conditions, then water will tend to move toward the surface to maintain equilibrium. The deeper into the surface the less this effect, meaning that unless conditions are extreme, poor curing is likely only to affect the outer 30 to 50 mm (Poole 2006).

The other potential cause of desiccation in a mixture may be that insufficient water was added to the mixture in the first place. Low w/cm mixtures are chosen to achieve improved performance, but below about 0.4 there is insufficient water to hydrate all of the cement (Addis and Alexander 1990). At very low values all of the water in the capillaries may be consumed causing them to contract, leading to so-called autogenous shrinkage. Making water available at the surface by ponding will only benefit the outer layer of the element and have no effect on the core. Severe desiccation may lead to progressive internal cracking and thus a loss in performance over time if the internal relative humidity drops below 80% within the first 7 days.

Many references discuss so-called initial curing, intermediate curing, and final curing (ACI Committee 308 2001; Poole 2006). The basic aim of these activities is to control moisture loss from the concrete surface, but different activities may be required at different times depending on finishing work still be conducted and the environment to which the concrete is exposed. Initial curing is normally efforts taken to prevent plastic shrinkage cracking occurring before the final finishing work is conducted. As such, any treatment must not compromise the quality of the concrete when finishing is carried out. This means that membrane curing compounds and flooding will not be effective at this stage, and evaporation retarders and fog sprays may be the preferred solution. Intermediate curing is sometimes required for surfaces that have been finished, but have not set or stopped bleeding. This may be the case in slabs on grade that are leveled and textured but not floated beyond the work needed to achieve final levels. In this case, similar approaches to those used for initial curing are used. Final curing refers to curing applied when finishing is complete or bleeding has stopped, and the concrete is sufficiently hydrated to resist water damage or marking, and can take the form of membrane-based curing compounds, flooding, or sheeting. Final finishing and curing should be delayed until after bleeding has ceased because water rising to the surface will tend to dilute the curing compound and will disrupt its ability to form a continuous membrane. Bleed water trapped or reworked back into the concrete surface will elevate the local w/cm leading to a soft, dusty, or scaling surface.

Evaporation retarders

A challenge to the construction of slabs on grade, particularly industrial floors, is that finishing work has to be delayed until bleeding has stopped and the mixture has achieved sufficient hardness. In drying conditions, including high temperature, low humidity, and high winds, the rate of bleeding may be slower than the rate of evaporation, resulting in the surface of the concrete drying out, leading to cracking (Figure 4.10; Poole 2006). Addition of water at this stage would effectively increase the local w/cm and decrease the long-term durability and polishing resistance of the mixture, and is therefore undesirable. Activities such as providing windbreaks and fog sprays may be helpful, but are often impractical in very exposed conditions or over large areas. If a fog spray is used, the water should not pond on the concrete surface.

Theoretically, one could seek to design a mixture that would balance the rate of bleeding with the rate of evaporation. However, the variability associated with the weather changing by the hour, and the undesirable consequences of having mixtures that bleed excessively means that this is not a realistic approach.

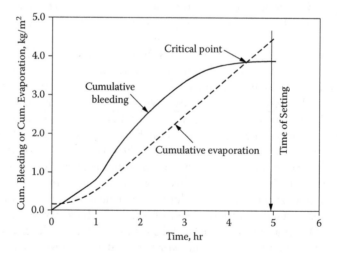

Figure 4.10 Plot comparing cumulative bleed and evaporation volumes. (From Poole, T. S., "Curing Portland Cement Concrete Pavements," FHWA-RD-02-099, Federal Highway Administration, McLean, VA, 2005.)

A good alternative to controlling the environment is to spray so-called evaporation retarders on the surface (Holt 2000; Liu et al. 2010). These are polyvinyl-alcohol-based compounds that are used to form an oil-like film on the surface of the concrete. Application rates are typically about 5 m²/L. They have the ability to reduce the rate of evaporation without compromising the w/cm. Use of these products is reportedly effective in reducing the risk of plastic shrinkage cracking, which is due to moisture loss from the surface before setting.

The guidance of the manufacturer should be followed with respect to application techniques and amounts. Typically they are sprayed on immediately after initial leveling is completed.

Poole (2006) reported that evaporation reducers provide benefit for a limited time, and as such might require repeated applications, depending on the rate of drying and the time of set. A reasonable practice would be to use evaporation reducers as a protection against excessive drying and to repeat applications whenever the sheen disappears until time of initial setting. Setting time can be monitored using thermal methods with transducers or thermocouples embedded in the concrete.

These products can be worked back into the surface with final finishing when used at recommended rates but should not be used as finishing aids.

Forms

An effective approach to limiting moisture loss from vertical elements is to leave the forms in place for an extended period, up to several days. The forms will help in preventing moisture loss, and if they are made of timber they will also act as thermal insulators. A difficulty associated with this practice is that it may limit the ability to proceed with the next lift. Rental costs of forms are normally high; therefore there is often pressure to recycle them as quickly as possible. Forms should not be removed when ambient temperatures are low to prevent thermal shock to the surface. If required, additional curing activities should be initiated as soon as possible after form removal to prevent drying out of the surface. This is important because once a surface has been allowed to dry, some carbonation may occur, densifying the surface making it difficult to get moisture back into the outer layer, even if hydration has not proceeded as far as desired.

An innovative approach to formwork that can influence concrete performance is the use of permeable formliners. These are geotextile-like products that are secured to the forms before the concrete is placed. They help in removing some water from the surface layer of the fresh concrete, effectively reducing the w/cm and so improving permeability (Taylor 1995; Torrent 2012). In effect then, the bulk concrete can be placed at a higher w/cm while the surface is still able to provide the element with the protective skin it needs. Mixture costs can therefore be reduced. The quality of the surface is also improved because bugholes are prevented and a more uniform attractive finish is easily obtained. It also means that inverted surfaces can be formed without air voids being trapped under the form. If these products are used, it is important that effective moisture control is provided as soon as the forms are removed. One approach is to leave the formliner in place, which can then be regularly wetted or covered with plastic sheeting.

Misting, flooding, and spraying

Misting is an effective means of protecting fresh concrete, particularly flat slabs, from premature cracking (Figure 4.11). As such it can be used for initial and final curing, but in practice it is normally only used in the initial stages of hydration. Mist is normally provided by means of fine nozzles such as used in garden irrigation systems. It should be started as soon as the concrete has been formed and leveled. The main aim is to raise the relative humidity of the air around the concrete and so slow the rate of drying. As such it is necessary to ensure that the mist is not blown away from the concrete, yet to prevent excess water from dropping on and marking the concrete surface. At a minimum the misting should continue until finishing activities are started.

The practice of flooding horizontal surfaces for the purposes of curing has been used for many years. In many ways the approach is an effective way of ensuring that the surface is provided adequate water for hydration. This will also help to maintain a uniform temperature, but ponding water should not be more than about 11°C (20°F) cooler than the concrete. Ponds

Figure 4.11 Mist sprays to increase RH above the surface of the concrete. (Portland Cement Association image 120-22A. With permission.)

can be formed by sand dikes around the perimeter but can only be placed after the concrete has set.

Limitations include the practical difficulties of providing an effective means of containing the water, the need to frequently replace water lost to evaporation in hot dry weather, and the limitations imposed on other construction activities. Failure to keep the surface continually wet defeats the object, because once drying has occurred it is much harder to force moisture back into the microstructure. Where the appearance of the concrete is important, the water used for curing by ponding or immersion must be free of substances that will stain or discolor the concrete. The material used for dikes may also discolor the concrete.

Similarly, spraying the concrete after it has set may be adequate, but in practice it is almost impossible to ensure constant wetting of the surface. In addition, there is some risk that contaminants in the water may cause staining of the surface.

Recent data by Hajibabee et al. (2012) has also shown that this approach may not be as effective as using curing compounds in reducing curling in slabs on grade.

Immersion in lime-saturated water is commonly used in the laboratory for curing concrete test specimens. The lime is added to the curing water to prevent leaching from the samples.

Burlap, plastic, and waterproof paper

Saturated absorbent materials, such as burlap or geotextiles, can be used to reduce moisture loss. Materials should be clean to minimize discoloration. Absorbent materials must be thoroughly wetted then placed as soon as the concrete has hardened sufficiently to prevent surface damage. Care should be taken to cover the entire surface including the edges of slabs. The coverings should be kept continuously moist.

It is often a good idea to cover absorbent materials with a plastic sheet or to use products that have an absorbent layer bonded to a sealant layer to reduce the rate of evaporation from the material. Periodically rewetting the fabric under the plastic before it dries out should be sufficient.

Plastic sheet materials, such as polyethylene film, or waterproof paper can be used alone or with absorbent materials to reduce moisture loss (Figure 4.12). Covering with polyethylene film can cause patchy discoloration, especially if the concrete contains calcium chloride and has been finished using steel troweling. This discoloration is more pronounced when the film becomes wrinkled. There are also occasional reports of a red/brown discoloration occurring in concrete under the folds of plastic sheeting used in hot weather (Figure 4.13). It is believed that this is related to movement of iron compounds when the concrete is allowed to dry. Concrete that is dried rapidly when curing is ended is less prone to discoloration than

Figure 4.12 Plastic sheeting to prevent drying. (Portland Cement Association image 42250. With permission.)

Figure 4.13 Buff-colored staining that occurred under folds in plastic sheeting.

slow-dried surfaces (Miller et al. 1998). Other means of curing should be used when uniform color is important.

Another disadvantage of using this approach is that access to the protected surfaces is limited, meaning that other construction activities may be hampered.

Polyethylene film and waterproof paper should meet the requirements of ASTM C171. White sheeting should be used for curing exterior concrete during hot weather to reflect the sun, but black sheeting can be used during cool weather or for interior locations. These specifications also set out the requirements for a sheet material comprising burlap impregnated on one side with white opaque polyethylene film.

Absorbent materials and plastic sheets must be secured at the edges to prevent the wind from tunneling under them, effectively increasing the rate of drying.

Drying shrinkage of high-performance concrete was assessed for different curing regimes including sealed, in water, and none (Hindy et al. 1994). Findings included that drying shrinkage was reduced with extended time under plastic sheeting, and that the size of the specimen had a large influence on shrinkage. The latter is not surprising because the larger the specimen the smaller the relative volume affected by drying conditions.

In hot, dry, windy conditions it may be cost effective to place or drag a frame covered with tarpaulin behind a paving train (Figure 4.14). This will provide a short time in which the fresh concrete is protected from the environment, mostly for the purposes of controlling plastic shrinkage cracking.

Figure 4.14 Tents used to protect young concrete from wind and sun.

Curing compounds

The term "curing compound" is the generic label commonly applied to the family of products more formally known as liquid membrane-forming compounds.

These products consist of compounds that are sprayed on a concrete surface as a liquid, that subsequently form a membrane and so reduce evaporation from the surface. As in the other approaches discussed above, the aim is to maintain the relative humidity of the concrete surface above 80% for (preferably) seven days in order that cement hydration will continue. In many ways, curing compounds provide the best available approach to balancing the needs of curing large surfaces of concrete with minimum negative impacts. Concretes cured with membrane-forming curing compounds have been shown to have better resistance to deicing salts than those cured with an external supply of water (Marchand et al. 1994).

These compounds are typically based on waxes, and resins emulsified in water or solvent. Pressures to reduce volatile organic compound (VOC) emissions to below 350 g/L are forcing manufacturers to focus on water-based systems. In comparative testing, a compound based on chlorinated rubber was found to be the most effective, followed by the solvent-based compounds, and the least effective was the water-based type (Wang et al. 1994). Newer products based on poly(alpha-methylstyrene) are proving to be particularly effective (Hajibabee et al. 2012; Vandenbossche 1999). Emulsified linseed oil cure/sealer compounds are also used for curing some concrete pavements.

American standards address a variety of different types and properties of curing compounds. Other required properties are also addressed by some local authorities, as discussed later.

Type

The type refers to the color of the compound. Type 1 materials are clear and primarily used for architectural applications, especially surfaces that are not intended to receive any other covering. Type 2 materials are white pigmented to increase reflectivity and so cooling, or to aid inspectors in confirming that sufficient material has been applied. Type 1 may be also supplied with a fugitive dye, meaning that the color fades out after a week's exposure to sunlight.

Class

The class refers to the active ingredient. Class A is unrestricted with the active ingredients often being wax or linseed-oil-based materials. The active ingredients of Class B materials are required to be resins.

Water retention

The purpose of using curing compounds is to retain sufficient water in the concrete (that is a relative humidity greater than 80%) that hydration may proceed until adequate physical performance has been achieved.

Whether a curing compound application retains sufficient water under a given set of climatic conditions is strongly affected by the impermeability of the compound and the continuity of the membrane, which in turn is governed by the surface texture and the amount of material applied.

Data developed by Poole (2006) show that water retention appears to have a perceptible effect on near-surface properties. However, scatter in the data is large. Therefore, even though the effects appear to be linear, it would be difficult to reliably distinguish a concrete that lost 0.3 kg/m² from one that lost 1.0 kg/m² of water using near-surface test methods. White and Husbands (1990) used ASTM C1151-91 to study the effects of different water-loss properties of curing compounds on the near-surface zone of mortars. They were unable to detect any deterioration in performance until losses exceeded 1 kg/m² at 3 days.

The test method commonly used (ASTM C156-11) for determining the inherent water retention properties of a given curing compound exhibits high variability when tested in different laboratories. This can lead to a high probability that a good compound is rejected or that an undesirable compound is accepted, neither of which is desirable.

There is little published information on how much water retention is required. The specified requirements for curing compounds vary from a high value of 0.55 kg/m² at 3 days (ASTM C309-11) to low values of about 0.25 kg/m² at 3 days (some state departments of transportation) and 0.31 kg/m² at 7 days (U.S. Army Corps of Engineers 1990). It is likely that these values were developed by comparing strengths of samples with various dosages of compound. An average of the results reported in the major publications is approximately 0.6 kg/m² (Poole 2006).

An alternative consideration may be that variable properties, poor precision of the test method, and errors in application rate may be addressed by applying a stringent water-loss requirement.

Another practical problem is that curing compounds that meet very low water-loss specifications may be difficult to work with. Performance is most easily controlled by adjusting solids content, which in turn will affect viscosity and so application reliability, possibly defeating the object of having the higher performance material. A double application of a higher viscosity material may be preferable.

Drying time

Drying time requirements are generally set at a maximum of 4 h when tested in a laboratory. However, drying time in the field can be a practical

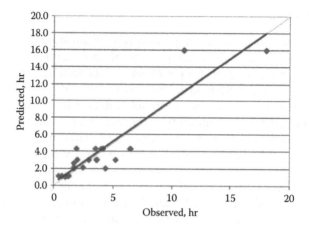

Figure 4.15 Comparison of calculated versus measured drying times for six curing compounds under a range of drying conditions.

problem, especially under adverse environmental conditions. This problem
is more marked when using low VOC curing compounds.

Poole (2006) demonstrated that drying time can be roughly expressed as
a function of evaporation rate as shown in Figure 4.15:

Drying time = 1.04 × (Evaporation rate)$^{-0.67}$

Dissipation

For some applications, such as when other coatings will be applied, it is
desirable that the curing compound be removed. Dissipating curing compounds are available that will break up and flake off when exposed to sunlight for a few days. Resin-based curing compounds will oxidize and begin
to wear off after about a month, depending on the amount of sun exposure
and traffic. Water or sand blasting may be also necessary. If drops of water
dropped on the surface form a bead, then the compound is likely still present (Dayton Superior 2011).

Volatile organic compounds (VOCs)

U.S. Environmental Protection Agency regulations limit VOC emissions to
350 grams per liter, particularly for indoor applications where accumulated
volatile organics can present a health hazard to workers. VOC-compliant
materials tend to be water based, which means their rate of drying could be
extended in humid conditions.

Work reported by Whiting and Snyder (2003) indicated that viscosities
of high VOC solvent-based (H) compounds were higher than the low VOC

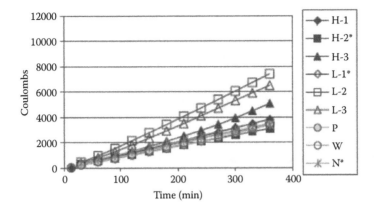

Figure 4.16 Rapid chloride penetrability test data for treatments after 28 days. (From Whiting, N. M., and Snyder, M. B., "Effectiveness of Portland Cement Concrete Curing Compounds," Transportation Research Record 1834, Paper No. 03-4014, 2003. With permission.)

water-based (L) compounds which may influence the ease with which it can be sprayed, and at the other extreme whether the product would run off vertical surfaces. In addition free water remaining on the surface of the specimen caused problems with some water-based compounds, resulting in nonuniform coverage and cracking in the membrane. Specimens treated with high VOC compounds tended to have less moisture loss, higher strengths, and lower permeability than the specimens treated with low VOC compounds. Specimens cured using plastic sheeting (P) retained more moisture and had better long-term strength development and lower permeability than any of the curing compounds (Figure 4.16).

Bond

Adhesives for floor covering materials generally do not adhere to surfaces covered with curing compound. The products should therefore be tested for compatibility, removed, or not used when bonding of overlying materials is necessary. Alternatively a resin-based compound may be used that will tend to dissipate after a few days.

Application method

Curing compounds should be applied using spray equipment immediately after final finishing of the concrete. The concrete surface should be damp, but bleeding should be effectively complete and the sheen should have evaporated before the product is applied (Figure 4.17). Complete coverage of the

Figure 4.17 Curing compound being sprayed by hand. (Portland Cement Association image 110-19. With permission.)

whole surface is critical. Sufficient material must be applied to account for the tendency of the material to run down into the texture or local low spots.

Power-driven spray equipment is recommended for uniform application on large surface areas. Spray nozzles and windshields should be set up to prevent wind-blown loss of product (Figure 4.18). If two coats are necessary

Figure 4.18 Curing compound being applied using a power spray system. Note the windshields attached to prevent wind losses. (Courtesy of Jim Grove.)

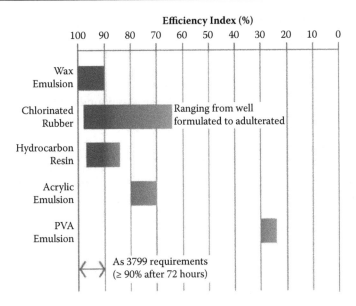

Figure 4.19 Australian guidance on the relative efficacy of different curing compound chemistries. (From Cement, Concrete, and Aggregates Australia [CCAA], "Curing of Concrete," 2006. With permission.)

to ensure complete coverage, then the second coat should be applied at right angles to the first.

If the compound is going to be applied to vertical surfaces or to deeply textured horizontal surfaces, then the viscosity must be high enough to prevent it from sagging. Such materials may not be suitable for spray-on applications because of the potential to clog spray nozzles and may have to be applied using rollers.

Composition

Australian guidance on the choice of compounds is based on the rankings shown in Figure 4.19 with the most effective and reliable curing compounds being those with a wax or resin base (Cement Concrete and Aggregates Australia [CCAA] 2006). Some U.S.-based authorities are requiring that curing compounds be comprised of poly(alpha-methylstyrene) because of its reportedly better performance.

Sealers

The primary purpose of a curing compound is to slow the loss of water from fresh concrete and it is applied immediately after finishing. Surface sealing compounds, however, are normally applied to hardened concrete in order to slow the penetration of harmful substances into the surface. They are therefore used to enhance durability, particularly in aggressive environments that even a well-proportioned mixture would be unable to resist.

Film-forming sealing compounds remain mostly on the surface with only a slight amount of the material penetrating the concrete because of the relatively large molecular structure of these compounds. Traffic and ultra-violet light will degrade the layer over time meaning that it will have to be reapplied periodically.

On the other hand, silane- and siloxane-based water-repellent penetrating sealers have a very small molecular size that allows them to penetrate and line the sides of the concrete pores. These sealers allow water vapor to evaporate from the concrete, while reducing liquid water penetration. Because the material is below the surface of the concrete, it is less sensitive to abrasion or environmental loading.

Penetrating sealers should only be applied to concrete that is clean and dry and at least 28 days old. The manufacturers' instructions will normally provide guidance on application rates and techniques.

There is also a class of materials known as "curing and sealing" compounds that are applied to fresh concrete. Curing compounds use a resin that breaks down in months, while curing and sealing compounds will resist sunlight and abrasion for years. They will therefore assist in preventing penetration by aggressive solutions, and so improve the potential durability of the structure. In addition, curing and sealing compounds can bond with paints and adhesives for flooring systems. Curing compounds with sealing properties are specified under ASTM C1315-11.

Internal curing

In some mixtures, particularly those with very low w/cm (below 0.40) there is a real risk of internal desiccation, because insufficient water is available to hydrate all of the cement in the mixture. This can lead to autogenous shrinkage, and in the extreme, a loss of performance over time.

As noted in the discussion earlier, most curing activities are applied to the surface of the concrete, and their effect is largely limited to a small zone near the surface. Therefore surface curing will not likely provide a significant benefit to mixtures prone to desiccation.

An innovative approach to addressing this is to provide reservoirs of water within the mixture that are not part of the initial mixing water; therefore it does not affect w/cm or capillary pore volumes. However, the water can be released at a later time to maintain sufficiently high RH in the system as hydration proceeds.

Two techniques area available to achieve this aim: the inclusion of saturated high-absorbent polymers or a lightweight fine aggregate that has a high porosity (Figure 4.20). Water absorbed in these materials does affect

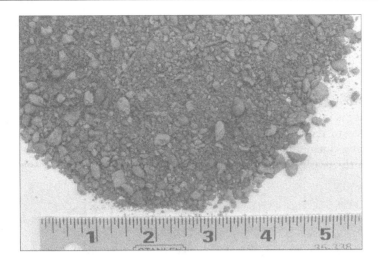

Figure 4.20 Porous lightweight fine aggregate used to provide internal curing. (Courtesy of Jiake Zhang.)

the w/cm but is available for hydration later in time. It is important that the amount of water absorbed in the material is sufficient for hydration, and that the rate of desorption from the material back to the paste is appropriate.

The advantage of using this approach is that the spacing between "reservoirs" is relatively small, meaning that all or most of the paste volume can be provided with the water it needs, locally. This significantly improves uniformity of the system (Figure 4.21).

Lightweight fine aggregate

The amount of lightweight aggregate (LWA) needed is relatively small, with only about 20% of the fine aggregate to be substituted with LWA, depending on the absorption of the LWA. This means that other properties of the mixture are not significantly affected. The target is to provide an extra 7% water in the aggregate as a mass of the cementitious materials (Bentz and Weiss 2010).

Bentz et al. (2005) developed an equation to determine the amount of lightweight fine aggregate needed for internal curing purposes:

$$M_{LWA} = \frac{C_f \times CS \times \alpha_{max}}{S \times \varnothing_{LWA}}$$

where
M_{LWA} = oven-dry weight of lightweight aggregate (lb)
C_f = cement content for the mixture (lb/yd^3)

CS = chemical shrinkage (lb of water/lb of cement) (Table 4.1)
α_{max} = maximum degree of cement hydration
S = degree of saturation of aggregate (0 to 1)
Φ_{LWA} = absorption of lightweight aggregate (lb water/lb dry LWA)

Reportedly cracking risk in thin slabs is significantly reduced, while per-meability is improved and drying shrinkage is reduced (Figure 4.22). All of this leads to reduced risk of cracking and greater longevity. Use of internal curing does not preclude the need for effective curing at the surface, because the two activities have different purposes.

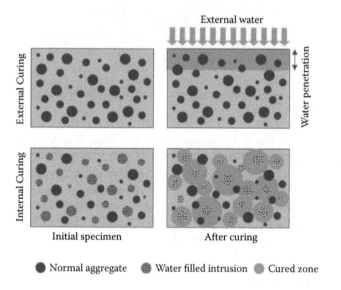

Figure 4.21 The distribution of curing water using different methodologies. (From Bentz, D. P., and Weiss, W. J., "Internal Curing: A 2010 State-of-the-Art Review," NISTIR 7765, National Institute of Standards and Technology, Gaithersburg, MD, 2010. With permission.)

Table 4.1 Coefficients for Chemical Shrinkage

Cement Phase	Coefficient (Pound of Water/Pound of Solid Cement Phase)
C_2S	0.0704
C_3S	0.0724
C_3A	0.115[a]
C_4AF	0.086[a]

Source: Bentz, D. P., Lura, P., and Roberts, J. W., *Concrete International*, 27, 2, 35–40, 2005. With permission.

[a] Assuming total conversion of the aluminate phases to monosulfate.

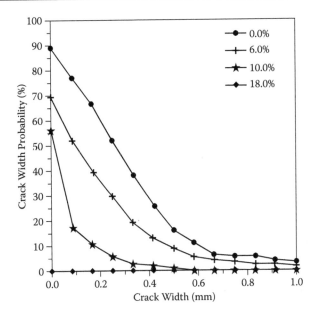

Figure 4.22 Probability of crack widths with different replacement. c = volumes of light-weight fine aggregate (LWA). (From Bentz, D. P., and Weiss, W. J., "Internal Curing: A 2010 State-of-the-Art Review," NISTIR 7765, National Institute of Standards and Technology, Gaithersburg, MD, 2010. With permission.)

Work at NIST by Bentz and Weiss (2010) showed that autogenous shrinkage of mortar samples was significantly lower in samples containing lightweight fine aggregate than systems containing lightweight coarse aggregate and conventional weight aggregate (Bentz 2010) (Figure 4.23).

Byard and Schindler (2010) conducted an extensive project investigating the effects of LWA as a partial sand replacement on factors affecting cracking. They reported the following findings:

- The modulus of elasticity is marginally reduced when lightweight aggregate is added to the mixture due to the reduced stiffness of the LWA.
- Internal curing (IC) mixtures have similar or slightly higher compressive and tensile strengths at all ages than the normal weight control concrete.
- Coefficient of thermal expansion was significantly reduced in internal curing mixtures.
- Thermal diffusivity was reduced in internal curing mixtures.
- The IC concretes experienced reduced autogenous shrinkage compared to the control concrete due to the fraction of fine LWA replacement and its reduced modulus of elasticity.

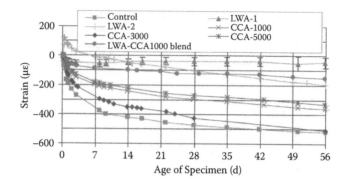

Figure 4.23 Autogenous deformation in microstrain for mortar mixtures with various lightweight aggregate systems. (From Bentz, D. P., and Weiss, W. J., "Internal Curing: A 2010 State-of-the-Art Review," NISTIR 7765, National Institute of Standards and Technology, Gaithersburg, MD, 2010. With permission.)

- Risk cracking at early ages in bridge deck concrete applications was modeled and found to be delayed with the use of internal curing.
- The temperature peaks of the Control and IC concretes are similar, but the time to reach the peak temperatures was slightly retarded in the internal curing mixtures.
- Data by Guthrie shows that moisture contents, measured using embedded capacitance-based sensors, are measurably higher in systems containing lightweight fine aggregate. Cracking in the decks was significantly lower in the IC deck than in the conventional.

Moisture contents of bridge decks were found to be measurably higher in the deck containing the internal curing aggregate (Figure 4.24; Guthrie and Yeade 2013). This correlated with fewer cracks observed in the decks.

Guidance on quality parameters for the aggregate and mix proportioning are available from the Expanded Shale, Clay, and Slate Institute (2012) based on a number of field test locations including bridge decks, pavements, and parking lots.

Superabsorbent polymers

Superabsorbent polymers (SAPs) provide an alternative means of internal curing. The most common SAPs used for internal curing are based on polyacrylamide, a thermoset polymer with chemical crosslinks that prevent dissolution in water (Siriwatwechakul et al. 2012a). The polymer may be coated with blast furnace slag or ground silica to prevent clumping. The unhydrated polymer is angular with particle sizes ranging from 50 to 300 µm. Absorption can be up to 2000 times the initial mass (Kevern and Farney 2012).

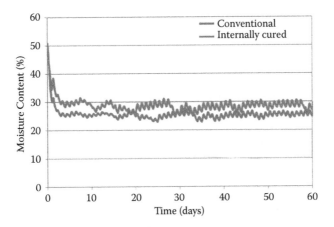

Figure 4.24 Average moisture contents for bridge decks with and without internal curing. (From Guthrie, W. S., and Yaede, J. M., "Internal Curing of Concrete Bridge Decks in Utah: Preliminary Evaluation," Paper No. 13-5374, Transportation Research Board, Annual Meeting, 2013. With permission.)

Results reported by Siriwatwechakul et al. (2010) indicate that the ability of SAPs to internally cure concrete depends on the method of polymerization (Figure 4.25). SAPs from bulk polymerization have demonstrated a superior strength gain compared to water-in-oil microemulsified SAPs in concrete exposed to drying conditions. Concrete with bulk polymerized SAPs show even greater strength gain when exposed to an external water source. Despite being chemically similar, the two types of SAPs appear to interact with pore solution differently resulting in different internal curing capabilities.

Figure 4.25 Scanning electron microscope (SEM) images of different forms of superabsorbent polymers that can be used for internal curing. (From Siriwatwechakul, W., Siramanont, J., and Vichit-Vadakan, W., *Journal of Materials in Civil Engineering*, 24, 8, 976–980, 2012a. With permission.)

Further work showed that solution-polymerized SAPs (Figure 4.25) appear to better regulate water transport needed for efficient hydration through an ion filtration effect, where cations electrostatically bond to the SAP structure. This effect causes absorption and desorption of water under saturated conditions when there is a change in the surrounding solution chemistry. This is relevant in early hydration of cement because there are constant changes in the pore solution chemistry. On the other hand, microemulsion polymerized SAPs appear to bind cations more strongly, leading to an observed decrease in desorption rates. Consequently hydration in concrete-containing micro-emulsion polymerized SAPs may be suppressed compared with concrete with no SAPs (Siriwatwechakul et al. 2012b).

Use of superabsorbent polymers is further discussed in a RILEM technical report (Mechtcherine and Reinhardt 2012).

Based on a study of concretes containing SAP (Wang et al. 2009):

- Water entrained by SAP is almost fully consumed within 7 days.
- Pores are left as the SAP releases water and so shrinks, leading to decreasing compressive strength with increasing SAP dosage.
- Use of prewetted SAP delays a drop in internal relative humidity.
- Increasing sap content reduces shrinkage (Figure 4.26).

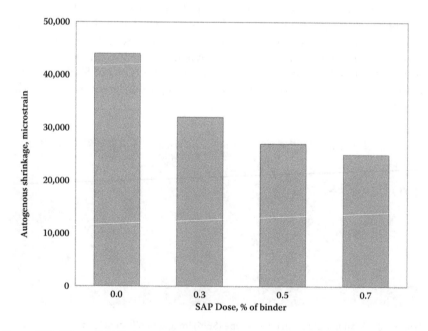

Figure 4.26 Effect of SAP dosages on autogenous shrinkage of concrete. (Data from Wang, F., Zhou, Y., Peng, B., Liu, Z., and Hu, S., ACI Materials Journal, 106, 2, 123–127, 2009. With permission.)

Lithium silicate sealant

A product has recently been made available that reportedly works somewhat like a curing compound, except that it does not form a membrane. The theory is that the silicates react with calcium hydroxide present as a by-product of cement hydration to create a form of calcium silicate hydrate. The new hydration product helps to densify the surface and reduce permeability and so moisture loss. Where the theory is inconsistent is that curing is needed to begin in the first few hours after hydration begins, and at this time it is unlikely that enough CaOH has formed to react with the lithium silicate.

Tawfig and Armaghani (2011) reported that lithium compounds reacted with free $Ca(OH)_2$ at the surface of the pavement to create a dense layer that helped to maintain the moisture in the concrete and so enhanced hydration at the surface. They also reported that use of lithium cure reduced the risk of plastic and drying shrinkage cracking in new slabs and decks.

Tamayo (2012) reviewed cracking in four bridge decks in Arkansas in which different curing methods were applied. On all three of the bridges evaluated the lithium-based compound outperformed the standard curing compound in terms of reduced cracking in the decks.

Rain protection

Although most of the discussion related to curing and moisture control is focused on making sure there is enough water available to the hydrating system, the effects of excess water on fresh concrete should not be ignored. Excess water on a flat surface is normally from two sources: applied by the finishing crew to aid surface preparation work, and rain.

If a rainstorm is likely to occur, then fresh slabs on grade should be covered. While a plastic sheet may result in marking of the surface, this will be preferable to loss in performance of the surface layer. Rainwater that does collect on the surface of concrete that has not set must not be worked back into the system but allowed to evaporate. The outer few millimeters may still be compromised and may have to be ground off to prevent dusting over time.

WHERE, WHEN, AND HOW LONG?

The discussion thus far in this chapter has focused on how various techniques can be used. This section looks at where and when it is necessary to cure concrete.

Where should curing be used? In principle, the answer to this question is simple: everywhere that a concrete surface is exposed, it should be protected from drying, and kept at a reasonable temperature.

When should it start? As soon as the concrete surface is exposed to the environment. The form of curing applied very early will depend on the circumstances, as discussed in later sections.

For how long? Again, in theory the answer is simple: long enough! The detail depends on the concrete mixture, the environment, and the demands to be placed on the structure. If it is assumed that little added benefit would be gained from continued hydration at the surface of a concrete once it is allowed to dry out, then the properties of the mixture should be at the required level before curing ends.

In addition, if the concrete is likely to be exposed to freezing conditions, sufficient time has to be allowed for some drying to occur, because saturated concrete that freezes will be damaged. The moisture content of the concrete should be below about 85% at the depth that freezing will occur if distress is to be avoided.

This question of when to end curing is moot if curing compounds are used, except that on slabs on grade, construction traffic may abrade the compound. Limited data are available to indicate the effective curing time provided by curing compounds, but one paper indicates that it is in the region of 4 to 7 days (Ballim et al. 1993).

The next chapter discusses the issues related to how much, and the following sections discuss in more detail the when and how long questions.

Protection between placing and curing

The time between placing concrete and application of curing compounds is critical for slabs under drying conditions because of the large area per unit volume and the high risk of shrinkage-related cracking. Conventional guidance is that evaporation rates exceeding 1.0 kg/m²/h are severe enough to warrant additional protection of the concrete.

This drying rate limit is apparently derived from typical bleeding rates of concrete (Al-Fadhala and Hover 2001). Holt (2000) published a plot of water-loss versus shrinkage, and tensile strain capacity versus time. These suggests that a water loss of about 1 kg/m² during the initial curing period might be sufficient to cause cracking.

If evaporation rates at the surface exceed the resupply of water through bleeding water rising from below, then the surface of the semirigid concrete will start to dry and shrink while it still has almost no tensile strength. Internal restraint from the aggregates will cause plastic shrinkage cracking. In particular, slipform paving mixtures tend to have low water contents and high content of fine materials, thus limiting the amount of bleeding and increasing the risk of plastic shrinkage cracking.

The rate and duration of bleeding primarily depends on the amount of fine materials in the mixture but is also affected by slab thickness and to a lesser extent the permeability of the support system. Evaporation rates can be assessed using the NRMCA nomograph (Figure 5.1) and weather records. Alternatively, evaporation rate can be determined by measuring mass loss from an open container of water.

Poole (2006) investigated the validity of the approach of tracking bleeding and evaporation by looking independently at bleeding rates, evaporation rates, and their related relationship with cracking risk. The following is based on the assumption that protective actions are required if evaporation exceeds bleeding.

Bleeding rate

The limits of acceptable evaporation used in ACI 308 of about 1 kg/m²/hr is based on past history of typical bleeding rates of concrete (Al-Fadhala and Hover 2001). Poole measured the bleeding rate for a number of concrete mixtures. Figure 4.27 shows average bleeding rates as a function of specimen thickness and w/cm. The relationship is roughly linear although the scatter is large.

Figure 4.28 shows data of bleed rate over time from a typical paving mixture (350 kg/m³ cement, w/c = 0.43). In this case the average bleeding rate was 0.19 kg/m²/h, with a peak of 0.33 kg/m²/h at about 2.5 h after mixing. It is notable that the average bleeding rate is considerably below the ACI guidance, even if adjusted for the thickness of the slab. In addition, the bleeding rate is not constant but builds to a maximum and drops again, meaning that considerable time is spent with bleeding rates well below the average, thus leaving the slab at risk of cracking for several hours.

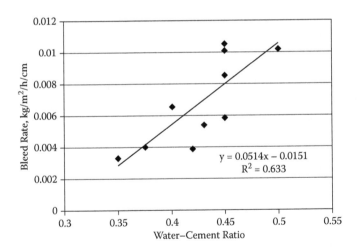

Figure 4.27 Plot of bleeding rate as a function of w/cm. (From Poole, T. S., "Curing Portland Cement Concrete Pavements, Volume II," FHWA-HRT-05-038, Federal Highway Administration, McLean, VA, 2006.)

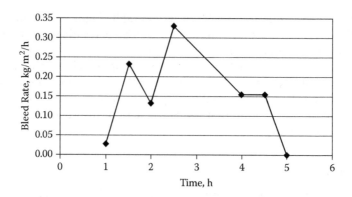

Figure 4.28 Plot of recorded bleeding rate over time. (From Poole, T. S., "Curing Portland Cement Concrete Pavements, Volume II," FHWA-HRT-05-038, Federal Highway Administration, McLean, VA, 2006.)

Evaporation rate

Al-Fadhala and Hover (2001) reviewed the accuracy of the National Ready Mixed Concrete Association (NRMCA) nomograph (Figure 5.1) and found that it reasonably represents the evaporation rate from a free-water surface for evaporation rates up to about 0.5 kg/m²/h. Above this, the nomograph tends to overestimate the rates of evaporation.

An effort was also made by Poole (2006) to verify the nomograph by measuring evaporation of bleed water from concrete and mortar under various controlled conditions. The results suggest that the mass loss of water from a surface is strongly dependent on wind flow patterns. Poole posited that when the wind is blowing directly onto the surface, aerosol formation potentially becomes a significant water transport mechanism, potentially doubling evaporation rates compared to those predicted by the nomograph.

Time of setting was also found to have a significant effect on losses of water from concrete. Testing by Poole (2006) showed that evaporation rates dropped dramatically after initial set as would be expected because the concrete microstructure is filling with hydration product, so limiting water movement.

Combined bleeding and evaporation

The conclusion from these studies is that maximum bleeding rates of paving mixtures may be less than 0.5 kg/m²/h. This, combined with the apparent underestimation of drying rates from the nomograph, means that a system following conventional guidance may still be at risk.

The estimated potential evaporation rates determined using the nomograph may therefore be useful in warning about potential risks, but decisions in the field should be based on what is being observed. For instance, if wind speeds are over about 2 m/s, then an extra level of caution is required. During placing, if surface sheen disappears within the first hour, or before final curing is applied, then that would be taken as evidence that water losses are high.

It should also be remembered that even if the system is drying out, the risk of cracking is dependent on the tensile strain capacity, that is, the stress induced by the shrinkage compared to the strength of the paste.

The most effective protection measures include use of evaporation retarders and misting, as discussed earlier.

Start of curing

For most applications, curing should start as soon as the concrete is exposed, because it is desirable to prevent the concrete from drying out. The exception to this is slabs on grade and pavements, because curing applied to a system that is still bleeding will be detrimental.

In the case of conventional concrete, the initial curing period starts immediately after placing and ends at about the time of initial setting, which is when final curing would be required. In the interim the concrete is allowed to stiffen, and finishing work will be conducted.

However in slipform pavement construction, finishing is essentially completed when the paving machine moves off the concrete except for a (hopefully) limited amount of touch-up work. Therefore, the initial curing period generally starts immediately after the concrete leaves the paving machine, and the need to start final curing is unclear. It is likely that often the concrete will not have reached initial set when final curing procedures are applied because the cure sprays are mounted on the same machine that forms the texture, which has to be done before the concrete sets. In addition, the sheen may disappear early because evaporation rates are greater than bleeding rates.

Bleeding starts shortly after the concrete is placed and ends by the time of setting. Being approximately tied to time of setting, loss of sheen may not occur for several hours, thus delaying application of curing compounds if the instruction is to delay application until after the sheen disappears. Application prior to the end of bleeding may cause the compound to bond with the surface of the specimen, if additional bleed water rises later it may delaminate a thin layer of surface mortar.

Poole (2006) attributed a case of delamination to the early application of curing compound under this mechanism. The pavement had been placed during windy drying conditions. A solvent-based curing compound was applied within about 15 minutes of placing with the aim of preventing plastic shrinkage cracks. About 3 months later, some small areas of delamination about 15 to 25 mm across were observed in the surface mortar. Petrographic examination showed cracks parallel to the surface at a depth of about 1 to 2 mm (see Figure 4.1). He also investigated this issue in the laboratory and concluded that if curing compound must be applied before bleeding ends and the sheen disappears, then it should be applied in two layers and not before the peak in bleeding rate.

Curing duration

Cessation of curing also depends on the type of curing being supplied.

The basic principle behind determining the duration of curing is that it can be ended when the required performance has been achieved in the mixture. Requirements in practice are often tied to strength but, as discussed in Chapter 2, strength in structural-sized elements is, in fact, only likely to be affected by moisture curing to a limited extent. Asselanis et al. (1989) reported that moist curing of high-strength concrete is normally unnecessary beyond 7 days, because after this time the concrete is effectively impervious.

The other common practice is to require a fixed time interval. For instance, AASHTO requirements are typical and call for 3-day curing for pavements and 7 days for structures, including bridge decks. The time for structures is increased to 10 days if more than 10% pozzolan is used or reduced to when 70% of specified strength is reached. ACI guidance applies fixed time periods as long as concrete temperatures are greater than 10°C. European Standard (EN) 206 accounts for strength gain of concrete, water–cement ratio, ambient temperature, and climatic conditions to give variable curing times, but maturity does not appear to be used. Australian guidance (Australian Standard 3600 1994) is based on durability for different climatic zones and environmental aggressiveness within the zones. Curing duration ranges from 3 days to 7 days, with some adjustments for accelerated curing.

None of the standard guidance mentions the effect of using pozzolan, even when such concretes typically gain strength more slowly. Poole (2006) quoted the literature as suggesting a minimum of 7 to 10 days when pozzolan is used.

Fixed time interval curing specifications tend to oversimplify the issue because rates of hydration vary widely between mixtures and in different environments. On the other hand, fixed time requirements are simple to measure and enforce. A better approach may be to rely on performance testing, either in situ strength assessed using thermal methods or preferably some other approach, such as abrasion resistance or permeability. Strength, however, is considered by some as a reasonable measure of degree of hydration of the cement because other properties develop at about the same rate as compressive strength. In situ strength development can be tracked using the maturity method. Temperature sensors have to be carefully located to avoid bias induced by being too close to free surfaces. An alternative is to use temperature-matched cylinders stored in a chamber that is controlled to the same temperature of the concrete as monitored using thermocouples.

Duration of moisture control

Concrete surfaces should be kept continually moist or protected from moisture loss until the surface has achieved the level of performance required at the surface.

Strength

Most commonly this is assessed based on strength despite the fact that curing is likely more normally required for permeability or abrasion properties than strength.

The European Standard EN 206-1:2000 for concrete states that curing time can be based on information on the assumed strength development of the concrete as shown in Table 4.2 or by a strength development curve at 20°C between 2 and 28 days.

Fixed period

Reportedly, most U.S. state departments of transportation call for moist curing between 3 and 4 days unless performance is demonstrated in situ (Poole 2006).

Parrott (1992) examined the effects on water absorption of moist curing conducted for various times. For the plain cement mixture, curing beyond 3 days only had a small effect, but the concretes with cement replacements had to be cured 28 days to achieve similar absorption values.

Hilsdorf (1995) developed a scheme to help select curing duration based on the type of (European) cement, w/c, and critical type of environment as shown in Table 4.3. The maximum is for CEM III systems (high supplementary cementitious material [SCM] contents) and even that value is 9 days.

Table 4.2 Strength Development of Concrete at 20°C

Strength Development	Estimate of Strength Ratio ($f\,cm,2/f\,cm,28$)
Rapid	0,5
Medium	0,3 to < 0,5
Slow	≥ 0,15 to < 0,3
Very slow	< 0,15

Source: European Standard (EN) 206-1, Concrete—Specification, Performance, Production, and Conformity, 2000. Reproduced with permission of CEN.

Table 4.3 Hildorf's Recommendations for Minimum Curing Periods

		Duration of Curing in Days for Different Cements				
		CEMI		CEM II-S	CEM III/A	CEM III/B
Criterion[a]	w/c	42.5R	32.5R	32.5R	42.5	32.5
C-Concept	0.4	<1	<1	<1	1.5	2
	0.5	<1	<1	<1	2	3
	0.6	<1	1	1.5	3	5.5
P-Concept	0.4	1	2	2	2	3
	0.5	1.5	3	3	3	5
	0.6	2.5	5.5	5.5	5.5	9
M-Concept	—	2.2	5.2	5.2	3.7	8
R1-Concept	0.4	1.5	2	3	2	4
	0.5	3	3.5	4	3.5	5
	0.6	7	7	7	7	7
R2-Concept	—	2	4	4	3	7

Source: Meeks, K.W., and Carino, N. J., "Curing of High-Performance Concrete: Report of the State-of-the-Art," NISTIR 6295, National Institute of Standards and Technology, 1999. With permission.

[a] C, carbonation; P, permeability; M, maturity; R, strength.

Performance tests

An alternative to a fixed time requirement is to use maturity, a method that can evaluate the strength of a mixture in the field. The method is based on measurements of the concrete temperature over time and comparing that with the previously established performance of the mixture. This is useful for estimating the in-place properties of the concrete and thereby being able to assess the need to continue curing, based on nondestructive test methods.

The basis for this method is that the strength of a concrete mixture is related to the degree of hydration, and thus the quantity of heat developed from the hydrating cement. A mix will therefore have the same strength at a given maturity no matter what conditions (time or temperature) occur. As shown in Figure 4.29, the area under the two curves is the same despite the different time temperature histories; therefore the degree of hydration of the mixture will be the same.

Each concrete mix has a unique strength–time relationship depending on the chemistry of the cementitious system. Maturity testing entails developing a curve that correlates the development of concrete properties for a specific concrete mix with time and temperature. Once the curve is developed, the concrete in the field can be estimated from in situ time and temperature data (Figure 4.29).

ASTM C1074 provides procedures for using modeled functions to develop a maturity curve. Data needs to be collected to characterize the

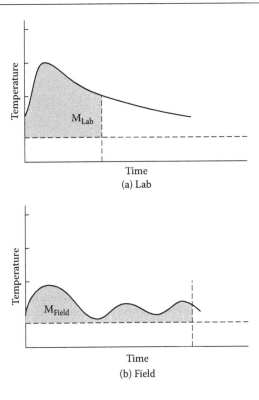

Figure 4.29 Maturity of the field concrete is equivalent to maturity of the laboratory concrete when the area under the curves is the same. (From Taylor, P. C., Kosmatka, S., and Voigt, J., eds., HIF-07-004, Ames, IA, National Concrete Pavement Technology Center, Iowa State University, Federal Highway Administration, 2006. With permission.)

mixture before placement in the field. Maturity in situ can then be calculated from temperature sensors embedded in the concrete.

Other approaches that may be considered include monitoring surface permeability or impact hammer rebound number (Figure 4.30 and Figure 4.31).

Ho (1992) reviewed sorptivity and carbonation rate results for vertical surfaces and observed that the methods could detect that practical curing procedures were equivalent to about 2 to 3 days of moist curing and that plastic sheeting and retention in formwork improved surface quality up to 7 days of protection. Similarly Torii and Kawamura (1994) showed that for high-strength concrete, 3 days of moist curing was sufficient for high resistance to chloride ion penetration and carbonation.

An alternative approach to quantifying application rates of curing compounds is to use reflectance to assess the degree of whiteness. The reflectance of a particular surface type can be calibrated against the application rate, and so tied back to expected performance of the system.

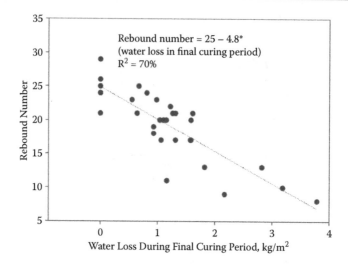

Figure 4.30 Rebound number of concrete surface versus water loss. (From Poole, T. S., "Curing Portland Cement Concrete Pavements, Volume II," FHWA-HRT-05-038, Federal Highway Administration, McLean, VA, 2006.)

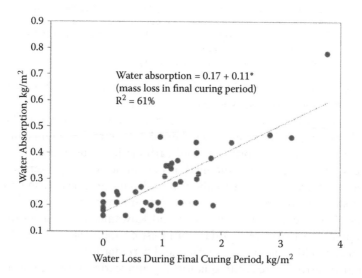

Figure 4.31 Surface water absorption versus water loss. (From Poole, T. S., "Curing Portland Cement Concrete Pavements, Volume II," FHWA-HRT-05-038, Federal Highway Administration, McLean, VA, 2006.)

REFERENCES

Addis, B. J., and Alexander, M. G., 1990, "A Method for Proportioning Trial Mixes for High Performance Concrete," High Strength Concrete, SP 121, American Concrete Institute, Farmington Hills, MI, pp. 286–308.

Al-Fadhala, M., and Hover, K. C., 2001, "Rapid Evaporation from Freshly Cast Concrete and the Gulf Environment," *Construction and Building Materials*, vol. 15, pp. 1–7.

American Association of State Highway and Transportation Officials (AASHTO), 1998, *Guide Specifications for Highway Construction*, 8th ed., AASHTO, Washington, D.C.

American Concrete Institute (ACI) Committee 207, 1996, "Mass Concrete" (ACI 207.1R-96), American Concrete Institute, Farmington Hills, MI.

American Concrete Institute (ACI) Committee 305, 1999, "Hot-Weather Concreting," ACI 305R-99, American Concrete Institute, Farmington Hills, MI.

American Concrete Institute (ACI) Committee 306, 1997, "Cold-Weather Concreting," ACI 306R-88, reapproved 1997, American Concrete Institute, Farmington Hills, MI.

American Concrete Institute (ACI) Committee 308, 2001, "Guide to Curing Concrete," ACI 308R-01, American Concrete Institute, Farmington Hills, MI.

ASTM C156-11, Standard Test Method for Water Loss (from a Mortar Specimen) through Liquid Membrane Forming Curing Compounds for Concrete.

ASTM C1074, Standard Practice for Estimating Concrete Strength by the Maturity Method.

ASTM C171-07, Standard Specification for Sheet Materials for Curing Concrete.

ASTM C309-11, Specification for Liquid Membrane-Forming Compounds for Curing Concrete.

ASTM C1151-91, Standard Test Method for Evaluating the Effectiveness of Materials for Curing Concrete (withdrawn 2000).

ASTM C1315-11, Standard Specification for Liquid Membrane-Forming Compounds Having Special Properties for Curing and Sealing Concrete.

ASTM D4263-83 (2012), Standard Test Method for Indicating Moisture in Concrete by the Plastic Sheet Method.

ASTM F1869-11, Standard Test Method for Measuring Moisture Vapor Emission Rate of Concrete Subfloor Using Anhydrous Calcium Chloride.

ASTM E1347-06 (2011), Standard Test Method for Color and Color-Difference Measurement by Tristimulus Colorimetry.

ASTM F2170-11, Standard Test Method for Determining Relative Humidity in Concrete Floor Slabs Using *in situ* Probes.

Asselanis, J. G., Aitcin, P. C., and Mehta, P. K., 1989, "Effect of Curing Conditions on the Compressive Strength and Elastic Modulus of Very High-Strength Concrete," *Cement, Concrete, and Aggregates*, vol. 11, pp. 80–83.

Australian Standard 3600, 1994, Section 19, "Material and Construction Requirements," pp. 28–30.

Ballim, Y., Taylor, P. C., and MacDonald, H. K., 1993, "A Preliminary Assessment of the Effectiveness of a Liquid Membrane-Forming Curing Compound," *Concrete Beton*, special issue 66, pp. 25–26.

Bentz, D. P., Lura, P., and Roberts, J. W., 2005, "Mixture Proportioning for Internal Curing," *Concrete International*, vol. 27, no. 2, pp. 35–40.

Bentz, D. P., and Weiss, W. J., 2010, "Internal Curing: A 2010 State-of-the- Art Review," NISTIR 7765, National Institute of Standards and Technology, Gaithersburg, MD.

Byard, B. E., and Schindler, A. K., 2010, "Cracking Tendency of Lightweight Concrete," Auburn University Research Report Submitted to the Expanded Shale, Clay, and Slate Institute.

Carrasquillo, P. M., and Carrasquillo, R. L., 1988, "Evaluation of the Use of Current Concrete Practice in the Production of High-Strength Concrete," *ACI Materials Journal*, vol. 85, no. 1, pp. 49–54.

Cement, Concrete and Aggregates Australia (CCAA), 2006, "Curing of Concrete."

Dayton Superior, 2011, "Guide to Curing: Technical Data Sheet."

European Standard (EN) 206, Concrete—Specification, Performance, Production and Conformity.

Expanded Shale, Clay, and Slate Institute (ESCSI), 2010, "Internal Curing," http://www.escsi.org/ContentPage.aspx?id=205&ekmensel=1b7c39fc_61_74_btn-link (accessed August 2012).

Farny, J. A., and Tarr, S. M., 2008, "Concrete Floors on Ground," EB075, Portland Cement Association, Skokie, IL.

Guthrie, W. S., and Yaede, J. M., 2013, "Internal Curing of Concrete Bridge Decks in Utah: Preliminary Evaluation," Paper No. 13-5374, Transportation Research Board, Annual Meeting.

Hajibabee, A., Ebisch, T., and Ley, T., 2012, "Impact of Curing Methods on Curling of Concrete Pavements," National Concrete Consortium, presentation at Oklahoma Meeting, http://www.cptechcenter.org/t2/SP2012_Presentations_Reports/01b%20Ley-Impact%20of%20Curing%20Methods.pdf (accessed August 2012).

Hindy, E. E., Miao, B., Chaallal, O., and Aitcin, P. C., 1994, "Drying Shrinkage of Ready-Mixed High-Performance Concrete," *ACI Materials Journal*, vol. 91, no. 3, pp. 300–305.

Hilsdorf, H. K., 1995, "Criteria for the Duration of Curing," in *Proceedings of the Adam Neville Symposium on Concrete Technology*, V. M. Malhotra, ed., Las Vegas, June 12, pp. 129–146.

HIPERPAV, http://www.hiperpav.com/ (accessed September 2012).

Ho, D. W. S., 1992, "The Effectiveness of Curing Techniques on the Quality of Concrete," Technical Report TR92/3, CSIRO, Australia.

Holt, E. E., 2000, "Where Did These Cracks Come From?" *Concrete International*, vol. 22, no. 9, pp. 57–60.

Kanare, H. M., 2008, "Concrete Floors and Moisture," EB119, Portland Cement Association, Skokie, IL.

Kevern, J. T., and Farney, C., 2012, "Reducing Curing Requirements for Pervious Concrete Using a Superabsorbent Polymer for Internal Curing," Transportation Research Board Annual Meeting 2012.

Kjellsen, K. O., Detwiler, R. J., and Gjorv, O. E., 1991, "Development of Microstructures in Plain Cement Pastes Hydrated at Different Temperatures," *Cement and Concrete Research*, vol. 21, pp. 179–189.

Kosmatka, S. H., and Wilson, M. L., 2011, *Design and Control of Concrete Mixtures*, 15th ed., Portland Cement Association, Skokie, IL.

Liu, J. P., Li, L., Miao, C. W., Tian, Q., Ran, Q. P., and Wang, Y. J., 2010, "Characterization of the Monolayers Prepared from Emulsions and Its Effect on Retardation of Water Evaporation on the Plastic Concrete Surface," *Colloids and Surfaces A: Physicochemical and Engineering Aspects*, vol. 366, pp. 208–212.

Marchand, J., Sellevold, E. J., and Pigeon, M., 1994, "The Deicer Salt Scaling Deterioration of Concrete—An Overview," Durability of Concrete, Third International Conference Nice, France, SP-145, V. M. Malhotra, Ed., American Concrete Institute, Farmington Hills, MI, pp. 1–46.

McCullough, B. F., and Rasmussen, R. O., 1999, "Fast-Track Paving: Concrete Temperature Control and Traffic Opening Criteria for Bonding Concrete Overlays, Volume II: HIPERPAV User's Manual," Publication No. FHWA-RD-98-168, Washington, D.C.

Mechtcherine, V., and Reinhardt, H. W., eds., 2012, "Application of Super Absorbent Polymers (SAP) in Concrete Construction: State-of-the-Art Report Prepared by Technical Committee 225-SAP," (RILEM State-of-the-Art Reports), Springer, Dordrecht.

Meeks, K. W., and Carino, N. J., 1999, "Curing of High-Performance Concrete: Report of the State-of-the-Art," NISTIR 6295, National Institute of Standards and Technology.

Miller, F. M. G., and Conway, T., 2003, "Use of Ground Granulated Blast Furnace Slag for Reduction of Expansion Due to Delayed Ettringite Formation," *Journal of Cement, Concrete and Aggregates*, vol. 25, no. 2.

Miller, F. M., Powers, L. J., and Taylor, P. C., 1998, "Investigation of Discoloration of Concrete Slabs," PCA R&D Serial No. 2228, Portland Cement Association, Skokie, IL.

Mindess, S., Young, J. F., and Darwin, D., 2003, *Concrete*, 2nd ed., Prentice Hall, Upper Saddle River, NJ.

Parrott, L. J., 1992, "Water Absorption in Cover Concrete," *Materials and Structures*, vol. 25, no. 149, pp. 284–292.

Poole, T. S., 2006, "Curing Portland Cement Concrete Pavements, Volume II," FHWA-HRT-05-038, Federal Highway Administration, McLean, VA.

Schindler, A. K., and McCullough, B. F., 2002, "The Importance of Concrete Temperature Control During Concrete Pavement Construction in Hot Weather Conditions," Transportation Research Board, Annual Meeting, Washington, D.C., January.

Senbetta, E., and Malchow, G., 1987, "Studies on Control of Durability of Concrete Through Proper Curing," Concrete Durability, Katharine and Bryant Mather International Conference, SP-100, J. M. Scanlon, ed., American Concrete Institute, Farmington Hills, MI, pp. 73–87.

Siriwatwechakul, W., Siramanont, J., and Vichit-Vadakan, W., 2010, "Superabsorbent Polymer Structures," International RILEM Conference on Use of Superabsorbent Polymers and Other New Additives in Concrete, Technical University of Denmark, Lyngby, Denmark, August 15–18.

Siriwatwechakul, W., Siramanont, J., and Vichit-Vadakan, W., 2012a, "Behavior of Superabsorbent Polymers in Calcium- and Sodium-Rich Solutions," *Journal of Materials in Civil Engineering*, vol. 24, no. 8, pp. 976–980.

Siriwatwechakul, W., Siramanont, J., and Vichit-Vadakan, W., 2012b, "Ion Filtration Effect of Superabsorbent Polymers for Internal Curing," Recent Advances in Concrete Technology and Sustainability Issues, Proceedings of 12th International Conference, Prague, October.

Tamayo, S., 2012, "Evaluation of High Performance Curing Compounds on Freshly Poured Bridge Decks," Arkansas State Highway and Transportation Department.

Tawfig, K., and Armaghani, J., 2011, "Evaluation of Shield-Forming Curing for Concrete Pavement," Florida State University, Tallahassee, FL.

Taylor, P. C., 1995, "An Assessment of the Benefits of Using Controlled Permeability Formliner for Concrete in South Africa," *Journal of the South African Institution of Civil Engineers*, vol. 37, no. 2, pp. 29–32.

Taylor, P. C., Kosmatka, S., and Voigt, J., eds. 2006, "Integrating Materials and Construction Practices for Concrete Pavements," HIF-07-004, National Concrete Pavement Technology Center, Iowa State University, Federal Highway Administration, Ames, IA.

Torii, K., and Kawamura, M., 1994, "Mechanical and Durability-Related Properties of High-Strength Concrete Containing Silica Fume," in *High-Performance Concrete, Proceedings of ACI International Conference*, SP-149, V. M. Malhotra, ed., American Concrete Institute, Farmington Hills, MI, pp. 461–474.

Torrent, R., 2012, "Technical-Economical Consequences of the Use of Controlled Permeable Formwork," Proceedings of the International Conference on Concrete Repair, Rehabilitation and Retrofitting Cape Town, South Africa, September 2–5, http://www.iccrrr.uct.ac.za/ (accessed September 2012).

U.S. Army Corps of Engineers, "Specifications for Membrane-Forming Compounds for Curing Concrete," CRD-C 300.

Vandenbossche, J. M., 1999, "A Review of the Curing Compounds and Application Techniques Used by the Minnesota Department of Transportation for Concrete Pavements," St. Paul, Minnesota, Minnesota Department of Transportation.

Wang, F., Zhou, Y., Peng, B., Liu, Z., and Hu, S., 2009, "Autogenous Shrinkage of Concrete with Super-Absorbent Polymer," *ACI Materials Journal*, vol. 106, no. 2, pp. 123–127.

Wang, J., Dhir, R. K., and Levitt, M., 1994, "Membrane Curing of Concrete: Moisture Loss," *Cement and Concrete Research*, vol. 24, no. 8, pp. 1463–1474.

White, C. L., and Husbands, T. B., 1990, "Effectiveness of Membrane-Forming Curing Compounds for Curing Concrete," Miscellaneous Paper SL-90-1, U.S. Department of Army, Waterways Experiment Station (now Engineer Research and Development Center), Vicksburg, MS.

Whiting, N. M., and Snyder, M. B., 2003, "Effectiveness of Portland Cement Concrete Curing Compounds, Transportation Research Record 1834, Paper No. 03-4014.

Chapter 5

Measurement and specifications

BALANCING THEORY AND PRACTICE

Many people agree in principle that curing is a good thing. But it often does not get done or done well. Why?

- It costs money and contractors don't always get paid for it, or if they do, it is not measured effectively
- Specifications do not address the details of curing adequately
- It takes time and effort and interferes with work progress
- It benefits the owner, not the contractor
- The benefits are not immediately apparent

An interview with an engineer recently elucidated the following statement: "We did not pay much attention to curing because we could get away with ignoring it."

So the burning issues are: Are the benefits worth the effort, and how do we ensure that the work is carried out adequately? It is apparent that some curing is beneficial and necessary; the art is providing enough to get the bulk of that benefit without compromising construction progress or letting it get out of balance.

This chapter addresses specifications and measurement issues relevant to curing.

HEATING AND COOLING

Minimum temperature

In principle, the aim of any curing activity applied to concrete being placed in cold conditions is to keep it warm enough to allow hydration to proceed. Lower temperatures will reduce hydration rates, meaning that it will take the mixture longer to achieve the required performance. This means that forms will have to be left in place longer and that the concrete should be protected from freeze–thaw cycles until it is able to resist them.

The Portland Cement Association (PCA) recommends that protection can be in the form of enclosures, windbreaks, portable heaters, insulated forms, or blankets (Kosmatka and Williams 2011).

American Concrete Institute (ACI) Committee 308.1–11 recommends that concrete be kept at a minimum of 10°C to allow hydration to proceed at a reasonable rate. This report also recommends that concrete that will be exposed to repeated freeze–thaw cycles must have a minimum compressive strength of 4500 psi.

ACI Committee 306 (1997) has developed a standard specification for cold weather concreting. The specification addresses the following points:

- Materials for protection should be at the site before concreting begins.
- Snow and ice must be removed from surfaces that the new concrete will contact.
- The subgrade and massive embedments must be thawed before placement.
- Placement temperatures are limited as shown in Table 5.1.
- Concrete must be kept moist for the specified period, after which it must be allowed to dry for 24 hours before freezing.
- Concrete may be exposed to one freeze–thaw cycle once it reaches 3.5 MPa (500 psi).

Table 5.1 Recommended Concrete Temperatures

		Section Size, Minimum Dimension			
Line	Air Temperature	<12 in. (300 mm)	12–36 in. (300–900 mm)	36–72 in. (900–1800 mm)	>72 in. (1800 mm)
Minimum Concrete Temperature as Placed and Maintained					
1	—	55°F (13 C)	50°F (10°C)	45°F (7°C)	40°F (5°C)
Minimum Concrete Temperature as Mixed for Indicated Air Temperature[a]					
2	Above 30°F (–1°C)	60°F (16°C)	55°F (13°C)	50°F (10°C)	45°F (7°C)
3	0 to 30°F (–18 to –1 C)	65°F (18°C)	60°F (16°C)	55°F (13°C)	50°F (10°C)
4	Below 0°F (–18°C)	70°F (21°C)	65 F (18 C)	60°F (16°C)	55°F (13°C)
Minimum Allowable Gradual Temperature Drop in First 24 hr after End of Protection					
5	—	50°F (28°C)	40°F (22°C)	30°F (17°C)	20°F (11°C)

[a] For colder weather a greater margin in temperature is provided between concrete as mixed and required minimum temperature of fresh concrete in place.

Source: ACI 306R-88 Cold Weather Concreting, ACI, Farmington Hills, MI, 2002.

- Differentials between the air and concrete surface should be less than 20°F (Poole 2006).
- Combustion heaters must be vented to outside the enclosure.
- Localized overheating and drying must be avoided.

Maximum temperature

As in the previous section, the aim of curing activities applied to concrete in hot conditions is to keep it cool enough to allow hydration to proceed at a controlled rate and to minimize the risk of thermal cracking. Higher temperatures will accelerate hydration leading to a potentially coarser, more permeable microstructure, while high differentials may lead to cracking. Internal temperatures above about 70°C increase the risk of delayed ettringite formation at a later time in some concrete systems.

Controlling maximum temperature is most efficiently achieved by controlling the mixture temperature rather than doing anything after the concrete has been placed. If necessary, light-colored protective layers such as plastic sheeting or curing compounds will help to reduce heat gain through the surface.

Temperature differentials between the core and the surface may be high enough to cause cracking, particularly if the nighttime temperatures drop significantly, in which case insulation layers are required.

Concrete that is exposed to high temperatures is likely to lose water through evaporation faster, meaning that greater effort has to be made in preventing moisture loss from the surface. Fog sprays, evaporation retarders, and curing compounds all help to reduce the rate of evaporation from slabs on grade.

PCA (Kosmatka and Wilson 2011) suggests that evaporation rates should be less 1 kg/m^2. The rate of evaporation can be estimated using a nomograph published by the National Ready Mixed Concrete Association (1960; Figure 5.1).

ACI Committee 305 has developed a standard specification for hot weather concreting. The specification includes the following points:

- The fresh concrete temperature should be limited to a maximum of 35°C.
- Temperature drops should be no greater than 22°C in the first 24 hours.
- Protective materials should comply with ASTM C171.

ASTM C171 covers the following:

- Daylight reflectance of white paper
- Tensile strength of curing paper
- Thickness and impact resistance of polystyrene sheet
- Composition of burlap/polystyrene sheeting

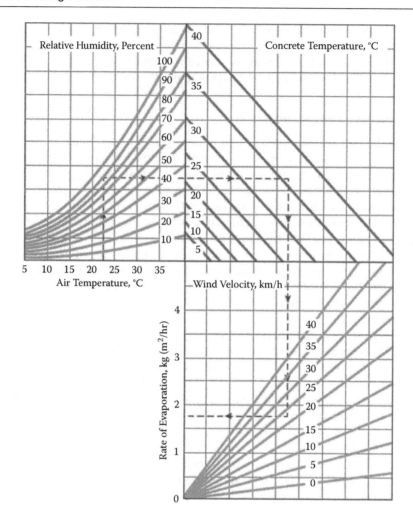

Figure 5.1 Nomograph to predict evaporation rates. (From Kosmatka, S. H., and Wilson, M. L., *Design and Control of Concrete Mixtures*, 15th ed., EB001, Portland Cement Association, Skokie, IL, 2011. With permission.)

Temperature differentials

The risk of cracking in concrete is increased when the difference in temperature between the interior and surface of a concrete element exceeds about 20°F. Methods that can be used to reduce this differential are discussed in ACI 308 and 207 and include the following:

- Minimize placement temperature
- Protect concrete surfaces when the air temperature is falling

- Use insulated forms or blankets
- Avoid removing forms when the air temperature is greater than 20°F more than the concrete temperature

MOISTURE CONTROL

The basic aim of moisture-related curing is to minimize moisture loss from the mixture until sufficient hydration has taken place. As discussed earlier and in ACI 308, ACI 301R, and European Standard (EN) 13670, this can be achieved by ponding, fog sprays, saturated/impervious sheeting, or curing compounds. Each is discussed in the following sections.

Ponding and fog sprays

If a surface is to be ponded, flooded, or sprayed, some guidelines on the quality of the water include the following:

- Temperature of curing water must be no more than 10°C cooler than the surface of the concrete.
- The water should not contain dissolved materials that will cause staining.
- If nonpotable water is used, care should be taken to protect workers in contact with it, and to prevent contamination of groundwater from the runoff.
- Water must be continually in contact with the concrete surface.
- Fog sprays should be adjusted to avoid water ponding on the surface of the concrete, especially before final set, to prevent marking the surface.

Burlap, plastic, and paper

Protective sheet materials should comply with ASTM C171, which requires water vapor transmission rate of less than 10 g/m^2.

In addition, if a surface is to be covered with absorbent materials:

- Absorbent materials must be nonstaining and free of sugar or fertilizer (ACI 308).
- Earth materials placed on slabs must be free of organic matter and free of particles larger than 25 mm.
- Straw or hay should be at least 150 mm thick (ACI 308).
- Burlap must meet (American Association of State Highway and Transportation Officials [AASHTO] M182).
- As with curing compounds, excessive evaporation must be avoided before application to avoid plastic shrinkage cracking. However,

additional strength may be required to prevent surface deformation if heavier mats are to be used.

- Moist curing inspection should be conducted once per shift or not less than two times per day, including nonworkdays.
- Low water–cement ratio concrete (~<0.40) will consume all of its mixing water in hydration, creating the potential for autogenous shrinkage. Added water-based curing should, in theory, be an effective tool for preventing this at the surface. However, water does not penetrate low water–cement ratio concrete after a relatively short time because the capillary pores become discontinuous in as little as 2 days in such mixtures. Any attempt to introduce water below the surface of the concrete through wet curing methods will be of limited benefit after this time.

Evaporation retarders

Evaporation retarders are not specified in ASTM at present. Users should follow the recommendations of reputable materials suppliers.

Curing compounds

Requirements for curing compounds are addressed in ASTM C309 and EN 14754. Factors addressed include the following:

- Whether the product has a fugitive dye, permanent pigment, or is clear
- Whether the product contains resin
- The product must form a continuous film when applied to damp concrete
- Maximum water loss is 0.55 kg/m^3 in 72 hours
- If required reflectance must not be less than 60%
- The product must dry to the touch in less than 4 hours

ASTM C309 also requires suppliers to comply with local volatile organic compound (VOC) limits, typically a maximum 350 grams per liter (g/L) of volatile solvents (Poole 2006). Reportedly, low VOC materials contain a relatively large amount of water, which can make the product slow drying and sensitive to environmental relative humidity (RH).

Curing and sealing materials are discussed in ASTM 1315. In addition to the parameters required in ASTM C309, the following are also required:

- Whether the product is permitted to "yellow" under ultraviolet light
- A minimum solids content of 25% or sufficient solids to form a 0.025 mm film when dry
- Maximum water loss is 0.40 kg/m^3 in 72 hours
- If required, reflectance must not be less than 65%

Water loss is measured using the method described in ASTM C156 in the United States or the European method EN 14754. Fresh mortar samples are prepared and covered with the curing compound at a fixed rate, normally 5 m²/L. The loss in mass is recorded over 72 hours and used to calculate the water loss taking into account the amount of volatiles in the product. The scatter on this test is large at 0.30 kg/m² in multilaboratory testing. This is a significant fraction of the 0.55 kg/m² limit imposed on curing compounds, meaning that there is a high probability that acceptable products are rejected and unacceptable products are accepted because of variability in the test.

It is likely that much of the between-laboratory precision problem is due to poorly detailed instructions in the test method as well as a lack of operator attention to the test method. The major variables include the following:

- Temperature
- Relative humidity
- Wind velocity
- Time of application of curing compound
- Surface finish of specimen

The method requires that specimens be placed in the curing cabinet after they are fabricated and left uncovered until the sheen disappears before applying curing compound. A difficult thing to standardize is the point at which the sheen on the test specimens disappears despite some supplemental instructions in ASTM C156.

Many specifications impose a minimum rate of application, often based on the 5 m²/L used for evaluating the product. However, other factors must be considered, including

- Loss of material in windy conditions
- Increase in effective surface area due to texturing or roughness; this may be a factor of 2 to 3 times the area obtained by simply multiplying length times width

Some specifications state that a second spraying may be required, perpendicular to the first coat, to ensure full coverage. One state department of transportation (DOT) reportedly requires that the final coverage must be as white as the engineer's hardhat. Figure 5.2 illustrates a surface that would not meet this requirement.

Application rate

Typical application rates vary among specifying agencies from about 2.5 m²/L to about 6 m²/L. Some agencies require application in two coats and the use of power spray equipment.

Figure 5.2 Roadway with nonuniform application of curing compound.

The application rate should take in to account the physical properties of the concrete surface, the properties of the curing compound, and the weather at the time of application. A better way to improve water retention may be to require more applications of a product containing lower solids contents rather than insist on tighter acceptance limits.

Poole demonstrated this concept using a curing compound with a relatively low solids content and a water retention of 0.49 kg/m². The product was applied at 0, 2.6, 5.4, and 9.7 m²/L. Typical results are shown in Figure 5.3 and demonstrate that a relatively high level of performance can be attained by increasing the application rate of a relatively poor curing compound applied in two coats. This is consistent with guidance for paints—a better finish is achieved with two coats than one thick one.

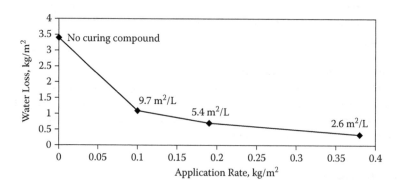

Figure 5.3 Water loss at 72 hours versus application rate of curing compound. (From Poole, T. S., "Curing Portland Cement Concrete Pavements, Volume II," FHWA-HRT-05-038, Federal Highway Administration, McLean, VA, 2006.)

Grooving and tining have the effect of increasing the effective surface area of the concrete. Shariat and Pant (1984) showed that moisture loss for a grooved specimen increased in proportion to the extra surface area created by the grooving. Tined and grooved surfaces may also contain vertical or near-vertical surfaces; therefore compounds that do not sag on vertical surfaces are recommended to avoid their running into the low spots of the tining.

To investigate this, three curing compounds were evaluated for their tendency to sag. Poole evaluated this property using a nonstandard test. Concrete specimens that were 140 by 120 mm were soaked in water for 1 hour, then surface-dried before they were turned on edge. Three different curing compounds with different viscosities were sprayed onto the vertical faces of the specimens at four application rates between 15 m^2/L to 5 m^2/L. Sagging was visually assessed after 5 minutes. The low viscosity material sagged at an application rate of 16.8 m^2/L, the intermediate viscosity material sagged between 4.6 and 6.7 m^2/L, and the high viscosity material did not sag at 5.0 m^2/L. It was recommended that lower viscosity materials should be applied in a number of coats.

Internal curing

No specifications have been developed yet for materials for use in internal curing applications. The Expanded Shale, Clay and Slate Institute (ESCSI 2012) has published a guide specification for aggregate-based internal curing. Suggested points to consider include the following:

- The aggregate should comply with the requirements for lightweight aggregate (ASTM C330).
- Smaller aggregate sizes are preferred to better distribute the available water through the microstructure.
- The aggregate must be prewetted before batching.
- The amount of lightweight aggregate can be calculated based on the chemical shrinkage of the mixture.

INTERNATIONAL SPECIFICATIONS

The following sections contain summaries of the requirements in various published or draft specifications for comparison. The extracts have been structured based on initial and final curing requirements as well as special cases where they are addressed. A thorough review is available in work conducted by Meeks and Carino in 1999, but some of these specifications discussed in this work are no longer published.

The documents discussed include ACI Committee 308 Guide Specification, Canadian Standards Association (CSA) A23, European Standard (EN) 206, and Unified Facilities Guide Specifications (UFGS) 03 39 00.

Initial curing period

The requirements given in ACI and CSA for initial protection include the following:

ACI 308

Initial curing should begin on concrete immediately after placement. Alternative approaches include use of fog sprays and evaporation retarders.

Fogging equipment should provide complete coverage to the affected area. Relative humidity above the slab should be high enough to prevent surface drying, yet without allowing surface water to stand on the surface. Treatment should be continuous until after final set.

Evaporation retarders can be used after strike off and between the floating operations to hold bleed water on the surface.

CSA

The exposed surface of high-strength concrete shall be kept moist through the use of a fog sprayer or other approaches applied immediately after initial finishing to reduce the risk of plastic shrinkage cracking.

Final curing period

All of the specifications referenced require provision of curing after setting. They take different approaches to discussing methods, materials, starting time, duration, and inspection.

Methods

Guidance on methodology includes the following:

ACI 308

The concrete should be protected from mechanical disturbance including traffic and running water.

Final curing should start with final finishing. Cycling wetting and drying is not permitted, regardless of the curing method.

If sheet material is used it should be placed as soon as possible without marring the surface. The sheeting should be continuous, placed beyond the edge of the concrete, and securely fastened or taped. Concrete under the sheeting should be continuously wet.

Curing compounds should be applied uniformly after final finishing and after the sheen has disappeared, using methods recommended by the manufacturer. The minimum rate should be based on manufacturer's recommendations or as stated in ASTM C309. The membrane should be protected from damage.

Flooding should be conducted by use of a dyke of sand or soil. Flooding should be continuous, starting as soon as possible without marring the surface, ensuring that water lost to evaporation or leakage is replaced.

Sprinkling should be conducted using lawn sprinklers or soaker hoses. Soaker hoses can be used to wet vertical surfaces still in their forms as long as no ersosion will occur. Surface must be kept continuously wet.

Fogging equipment should provide complete coverage to the affected area. RH above the slab should be high enough to prevent surface drying, yet without allowing surface water to stand on the surface. Treatment should be continuous until after final set.

Absorbent materials should be nonstaining and should be applied as soon as possible without marring the surface. Water should be applied as needed to keep the surface continuously wet.

EN 13670—Key points in European practice include the following:
- Curing is required to:
  - Minimize plastic shrinkage
  - Ensure adequate surface strength and durability
  - Protect the concrete from harmful weather including freezing
  - Protect the concrete from vibration and impact
- Curing can be provided by means of:
  - Maintaining air humidity >85%
  - Keeping the forms in place
  - Covering with waterproof sheets or covers
  - Applying water
  - Using of curing compounds
- Curing shall be implemented without delay.
- Curing compounds are not permitted on construction joints or surfaces where bonding is required.
- Concrete surface shall not fall below 0°C until a strength of 5 MPA is achieved.
- Maximum temperature shall not exceed 70°C unless performance of the mixture can be demonstrated to be satisfactory.

UFGS

Concrete should be kept moist and between 10°C and 30°C. The concrete should be protected from rapid temperature changes, mechanical damage, and flowing water. Temperature changes are limited to a maximum change of 3°C per hour.

Protection should begin as soon as free water disappears from the surface and should be provided for a minimum of 10 days. Moist curing should be provided by any of the following methods:

- Surfaces should be covered with a double layer of continuously wet burlap covered with 0.1 mm thick plastic sheeting. Burlap should be clean. No traffic is permitted while sheeting is in place.
- Ponding and sprinkling should be continuous.
- Curing compound should be applied to vertical surfaces as soon as forms are removed and to horizontal surfaces as soon as free water has disappeared. Application should be in two coats using spray equipment with a minimum pressure of 500 kPa at a uniform rate of 10 m²/L. Application of the second coat should be perpendicular to the first. The membrane should be protected from rain, traffic, or disruption for the entire curing period.

CSA

Curing should begin as soon as possible without marring the surface. Moist curing should be provided by any of the following methods:

- Ponding or continuous sprinkling
- Using an absorptive mat or fabric kept continuously wet
- Using damp sand, earth, or similar moist materials
- Applying curing compounds
- Covering with waterproofing paper or plastic film
- Using a vapor mist bath
- Keeping forms in contact with concrete surface
- Using other moisture-retaining methods

State DOTs—Several U.S. state DOTs also provided language from their specifications.

California

- Apply curing compound at a nominal rate of 150 sq ft/gal.
- At any point, the application rate must be within 50 sq ft/gal of the nominal rate.
- Apply the curing compound so that there are no runs, sags, thin areas, skips, or holidays.
- Apply the curing compound using power-operated spraying equipment.
- Apply the curing compound to the concrete after finishing the surface, before the moisture sheen disappears from the concrete surface, and before cracking.
- If the curing compound membrane is damaged before 7 days for structures or 72 hours for pavement, repair it with additional compound.

Colorado
- The concrete should not be left exposed for more than 30 minutes before being covered with curing compound.
- Curing compound should be applied by mechanical sprayers at a minimum rate of 1 gallon per 150 square feet of pavement surface.

Louisiana
- Spray exposed pavement surfaces with curing compound as soon as surface bleed water evaporates, or within one-half hour under drying condition.
- Do not apply curing compound during rainfall or to surfaces with standing water.
- Maintain curing continuously for 72 hours.
- Apply one coat of curing compound on nongrooved surfaces and two coats on grooved surfaces. Do not allow more than 2 hours between coats.
- Curing compound should be applied by mechanical sprayers at a minimum rate of 1 gallon per 100 square feet total application.
- Hand spraying is allowed on small irregular widths or shapes, and on surfaces exposed by form removal.
- After application of curing compound, pavement surfaces should have a uniform appearance of a blank white sheet of paper.
- Texture the plastic concrete before applying curing compound unless curing is being delayed.
- If texturing is not complete before placing curing compound, complete the surface texturing on the hardened concrete.

Minnesota
Minnesota DOT conducted a comprehensive review of curing practices and developed the following recommendations for low (<0.40) w/cm pavements (Vandenbossche 1999):
- Reflectance—Min.: 65% in 72 hours
- Water loss—Max.: 0.15 kg/m2 in 24 hours
- Water loss—Max.: 0.40 kg/m2 in 72 hours
- Settling test—Max.: 2 ml per 100 ml in 72 hours
- VOC content—Max.: 350 g/L (700 g/L for a cure and seal)
- Infrared spectrum of vehicle—100% poly(alpha-methylstyrene)
- Shelf life—Greater of 6 months or the calendar year the product was produced

This approach was found to increase material costs of curing compound by about 3 times, but was considered a small investment when compared to the total project cost and the potential benefit to the potential longevity of the pavement.

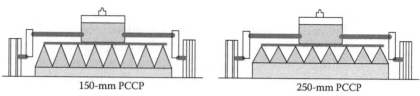

150-mm PCCP	250-mm PCCP
(a) Nozzle heights adjusted to obtain 30% overlap of adjacent spray patterns.	(b) Nozzle must be raised to retain 30% overlap for the 250-mm PCCP.

Figure 5.4 Illustration of required overlap of spray patterns. (From Vandenbossche, J. M., "A Review of the Curing Compounds and Application Techniques Used by the Minnesota Department of Transportation for Concrete Pavements," Minnesota Department of Transportation, St. Paul, MN, 1999. With permission.)

Curing compound must also be applied in such a manner that a uniform coverage is obtained and the loss of curing compound due to wind action is reduced:
 – Nozzle type chosen to produce a droplet large enough to prevent drift but small enough to maintain uniform coverage. Droplets with a diameter smaller than 200 microns (8 mils) are considered potential drift contributors
 – Nozzle spacing and boom height to insure a 30% overlap (Figure 5.4)
 – Nozzle orientation and cart speed to ensure a coverage rate of 4 m2/L
 – Wind shield to prevent loss due to wind
Texas
 – Apply 2 coats of curing compound
 – Each coat should not exceed 180 sf/gal
 – First coat should be applied within 10 minutes of texturing and the second coat within 30 minutes of texturing.
Many states also had language emphasizing the need to ensure that curing compound is well agitated before it is sprayed because it has a tendency to settle.

Materials

Requirements for the materials to be used for curing include the following:

ACI 308
 • Curing compounds should comply with ASTM C309 or ASTM C1315.
 • Sheet material should comply with ASTM C171.
 • Water used for curing should be free of staining or corrosive materials.

- Earth materials should be free of organic matter and particles larger than 1 inch.
- Burlap should comply with AASHTO M182.

EN 13670

The principle of the standard is that curing must continue until 50% of the characteristic strength is obtained. Requirements for the period of curing are based on time and strength, taking into account the exposure of the concrete and the rate of hydration of the mixture.

Exposures are classified on the basis of the exposure including the level of severity to carbonation, seawater, chlorides, freezing and thawing, and chemicals. Concrete that is exposed to a benign environment need only be cured for 12 hours. Other exposures are required to reach 50% of strength, or if wear and abrasion is likely, 70%.

The strength is assessed by in situ measurement, including techniques such as maturity or temperature-matched curing. An alternative is to use a table to set curing times based on the ratio of 2- to 28-day strengths (r) of the mixture, where $r = 0.5$ is a mixture that gains strength rapidly and $r = 0.15$ is slow. Typical plain portland cements produce mixtures with an r of about 0.3. Times in the table are doubled if the concrete is exposed to abrasion.

Other qualifications include the following:
- Added time is required if setting is later than 5 hours.
- Time below 5°C is not included in the curing time.
- Added protection is required for mixtures exposed to temperatures below 0°C before it reaches a strength of 10 MPa.

CSA

CSA requirements are similar to those listed in ACI 308.

Start of curing

All authorities agree that curing should begin as soon as possible without marring the concrete surface. This may mean that on horizontal surfaces final curing may have to be delayed until bleeding is complete and the sheen has disappeared.

Duration

There is some variation in the required duration of curing, but most authorities use a variation of strength or a fixed time.

ACI 308

Concrete should be cured for:
- At least 7 days as long as the concrete surface temperature is a minimum of 50°F

- Compressive strength is at least 70% of design strength based on:
  - Approved nondestructive test methods calibrated with compression tests
  - Compression tests on field cured samples
  - Maturity in accordance with ASTM C1074

CSA

Concrete should be cured for either 3 days at a minimum temperature of 10°C or for long enough to reach 40% of the specified 28-day compressive strength.

Concrete exposed to severe conditions, abrasion, or air pollution should be cured for an extra 4 days at a minimum of 10°C or until it reaches 70% of the specified 28-day strength.

UFGS

Concrete should be cured for a period governed by the cement type:
- Type III and plain cement with accelerator, 3 days
- Type I, IS, IP, or with silica fume, 7 days
- Type II or blends with less than 25% supplementary cementitious material (SCM), 14 days
- Blends with more than 25% SCM, 21 days

Inspection

Only the military specification addresses techniques to ensure that curing has indeed been carried out.

UFGS
- Inspect moist curing at least once per shift or twice per day every day.
- If the concrete has been allowed to dry out, the period should be extended by 1 day.
- Use the volume of curing compound used and the area covered to assess coverage rates. Inspect the surface to assess whether coverage is uniform.
- If coverage is inadequate, then the entire surface should be resprayed.
- Inspect sheets, laps, and taping at least once per shift or twice per day every day.
- If sheets are damaged, repair them and extend the curing period by one day.

Special conditions

Hot weather

ACI and CSA specifications address protecting concrete during hot weather.

ACI 308

Additional protection is required during hot weather. Initial curing techniques should be applied when evaporation rates are high.

- Shade formwork and concrete
- Provide windbreaks
- Use evaporation retarders or fogging
- Work at night

CSA

When the air temperature is at or above 27°C:

- Shade formwork and concrete
- Use a water spray or saturated absorptive fabric
- Concrete maximum temperature differentials should be limited to between 12°C and 29°C, depending on concrete thickness

When surface moisture evaporation exceeds 0.75 kg/(m²·h), windbreaks are required.

When surface moisture evaporation exceeds 1.0 kg/(m²·h):

- Dampen the subgrade before placing concrete
- Erect sunshades
- Lower the concrete temperature
- Cover the concrete surface with white polyethylene sheeting between finishing operations
- Apply fog spray immediately after placement and before finishing
- Begin the final curing immediately after trowelling
- Work at night

Cold weather

ACI 308

Measures to protect concrete in cold weather include the following:

- Do not expose concrete to exhaust fumes.
- Do not use added water methods if it is likely to freeze.
- Protect the concrete from freezing until it has reached 3500 psi compressive strength.
- After water curing is ended, protect the concrete from freezing for 3 days.
- Water retention methods may be used if the surface can be insulated or heated to above freezing.
- Maximum cooling rates are
  - 50°F per day for sections less than 12 inches in the least dimension.
  - 40°F per day for sections from 12 to 36 inches in the least dimension.

- 30°F per day for sections from 36 to 72 inches in the least dimension.
- 20°F per day sections greater than 72 inches in the least dimension.

CSA

Measures to protect concrete in cold weather include the following:

- Water curing of concrete should be terminated 12 hours before the end of the protection period.
- Concrete should not be placed on snow and ice, nor surfaces treated with calcium chloride.
- Protection should be provided by means of heated enclosures, coverings, insulation, or a suitable combination.
- Concrete surfaces should be protected from exposure to combustion gases or direct drying from heaters.
- For high-performance concrete, the maximum temperature differential should be 20°C.

UFGS

Measures to protect concrete in cold weather include:

- Maintain minimum temperature of 10°C for the first 3 days, followed by 0°C for the remainder of the curing period.
- Vent exhaust fumes from combustion heating units.
- Heat uniformly.

VERIFICATION OF CURING

Initial curing period

The greatest risk associated with inadequate protection of a concrete between placement and final set is that of plastic shrinkage cracking. This is particularly marked in concrete slabs on the ground but can be seen in the tops of formed concrete in structural elements. The consequences depend on the type of structure, with little remedial action needed in some vertical elements. Plastic shrinkage cracking in slabs on the ground is unsightly and may lead to reduced serviceability of the system.

Procedures to verify adequate protection are largely based on looking for cracking. Specifications tend to either ignore the issue or impose repair requirements when cracking occurs. Specifications should direct that if cracking starts to appear during this period, then measures to reduce evaporation or to compensate the concrete for water lost to evaporation must be implemented.

It would be desirable to develop a method that would detect deficiencies before damage starts to develop. One approach may be to monitor the surface sheen because loss of sheen is an indication that evaporation is exceeding bleeding increasing the risk of plastic shrinkage cracking.

Final curing period

Verifying water added and methods that use waterproof sheets are mostly a matter of regular inspections looking for dry spots or unprotected areas. Inspections should also address the risk of wind tunneling under sheet materials.

Measurement of curing compounds

Existing methods for estimating how much curing compound has been applied generally involve indirect measurements such as recording the amount of materials used and the surface area covered, or measuring flow rates and rate of movement of the application equipment. Both of these procedures can give good estimates of average coverage but will not detect localized irregularities.

Three methods were investigated by Poole (2001) to assess the effects of local irregularities in curing compound application:

- Mass
- Infrared
- Visible light reflectance

Direct measurement by weighing samples placed in the path of the spray is plausible. At typical application rates a 10-by-10 cm coupon would then show a change in mass of about 2.0 g. However, a significant error can be introduced by the evaporation of solvent, particularly in severe drying conditions. Up to 0.8 g could evaporate in 5 min meaning that attention is needed to collect and protect quickly. An alternative is to measure dry mass, but then a larger sample would be required to achieve sufficient precision, because most of the mass of a curing compound is solvent. This is the only method that will work on nonpigmented curing compounds.

The concept behind an infrared-based technique is that concrete with a poorly formed membrane will allow more water to evaporate than one with a well-formed membrane. The difference in evaporation should therefore result in a temperature difference that can be detected using an infrared camera. Poole (2006) assessed the plausibility of this and found that the temperature of coated specimens rose above the temperature of uncoated specimen almost as soon as the curing compound had dried but persisted only for a short while. This technique does not appear to have promise for general field use, given the potential difficulties of calibrating a signal because of a number of complicating effects including sunlight, wind, and evaporation rates between mixtures.

Visual examination is a common but subjective method for verifying application of white-pigmented curing compound. Poole (2006) investigated a means to quantify the practice using fresh mortar and black paper specimens coated with different levels of a white-pigmented curing compound. Curing compound was applied using a paint sprayer at variable application rates and used as standard reference set. Other specimens were coated with variable amounts of curing compound and treated as unknowns. An estimate of the application rates was made by visual comparison with the standards and was within about 2 m²/L of the actual application rate. The samples were also measured using a portable reflectometer in accordance with ASTM E1347 and showed a good correlation (Figure 5.5). A drawback to this approach is that it has to be calibrated for every mixture.

Performance-based methods

As noted earlier, the test method used to evaluate water loss for curing compounds is subject to a large experimental error. This means that there is a high risk of a false signal being recorded in any given laboratory test of a product. This is undesirable in an age of tight economics and a desire to use products efficiently for sustainability reasons. In addition, performance of the concrete is also affected by application rates, which are not easy to determine.

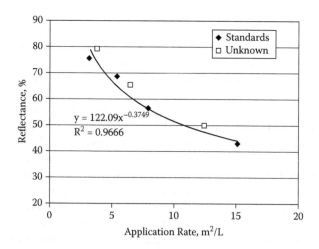

Figure 5.5 Reflectance versus application rate for white-pigmented curing compound applied to fresh mortar at different rates. (From Poole, T. S., "Curing Portland Cement Concrete Pavements, Volume II," FHWA-HRT-05-038, Federal Highway Administration, McLean, VA, 2006.)

The alternative then is to use tests conducted on the surface of the concrete as placed and cured, to confirm that adequate curing has been provided.

Strength is a common measure of the overall state of hydration of a concrete sample, but does not indicate the effects of moisture loss in the surface zone. Several test methods have been investigated to assess their ability to indicate the degree of property development near the surface as a consequence of curing. Many of these require specialized equipment and are not easily conducted in the field. Poole (2006) conducted some work to investigate some of these approaches.

A number of methods rely on measuring strength, both directly or by use of nondestructive methods, or on measuring surface or near-surface physical properties. Strength methods are covered in ACI 228 and include in-place curing of test cylinders, ultrasonic pulse velocity, rebound hammer, pull out, and penetration resistance. Other methods include measurement of permeability, water absorption, relative humidity, abrasion resistance, and hardness (Kropp and Hilsdorf 1995).

A common problem with in situ tests is the large effect that the moisture condition of the concrete will play on the result. Most of these methods are most reliable when test specimens are prepared and analyzed in the laboratory after drying. Some of the field methods do include moisture correction procedures. The simplest test found was based on a relative humidity button containing a moisture-sensitive dye, allowing visual verification that the surface of the concrete was wet (Carrier and Cady 1970), although it is not widely used (Senbetta 1994).

Other methods described in ACI 228 potentially could be adapted for measuring changes in near-surface properties as an indicator of curing.

Gas permeability

The approach of determining gas permeability of the surface layer was suggested by Perrie (1994) as one way of addressing this need (Figure 5.6).

The idea is that surface performance of the mixture using sorption or permeability can be calibrated based on the amount of water soaking provided. Field testing of the in situ concrete can then be used to assess the quality of the curing, and this can be used as the basis of payment for a line item on curing in the bill of quantities. The greatest difficulty with this approach is that variability of field performance tests is also still high, but the concept warrants further investigation. Alternative methods are discussed next.

Ultrasonic pulse velocity

The speed of sound is a direct function of the modulus of the elasticity (E) of the material it is traveling through. Based on this, a test using ultrasonic

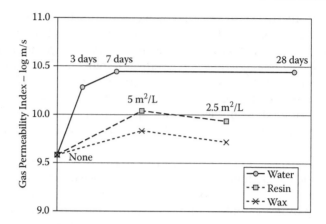

Figure 5.6 How permeability of concrete treated with different types and amounts of curing compounds can be correlated with an equivalent wet curing period. (Data from Perrie, B. D., "The Testing of Curing Compounds for Concrete," master's dissertation, University of the Witwatersrand, Johannesburg, South Africa, 1994.)

pulse velocity (UPV) can be used to monitor the time it takes for a sound signal to travel through a sample and so correlate it to E, and so the degree of hydration. A barrier to using this approach is that the equipment required is specialized. Equipment available comprises a pair of sensors, one that generates a sound spike and the other that listens for it. It is necessary to ensure that a suitable couplant such as petroleum jelly is used to ensure good contact between the sensors and the concrete surface. This means that sensors do have to be on opposite faces of a sample. A processor is required to generate the sound and to monitor and report the time interval between delivery and arrival of the signal. Such a device can also be used to monitor time of set of a mixture.

Tests by Poole (2006) showed that an uncured surface could be detected, but there was insufficient difference in UPV to differentiate between different application rates of curing compound. More sophisticated hardware and data analysis procedures may still be able to detect such differences.

Rebound hammer

The rebound hammer (ASTM C805) is primarily intended to provide an in situ estimation of strength, and is most commonly used as a comparative tool to compare known and unknown portions of a structure, or to assess variability.

The method is sensitive to near-surface properties of the concrete and may not represent the properties of the concrete as a whole. This effect may be a benefit when assessing the changes in surface properties as a consequence of curing practice.

Tests have shown a relatively strong correlation between water absorption and rebound number, as well as between rebound number and water lost in the final curing period (see Chapter 4, Figure 4.30; Poole 2006). Rebound numbers may be useful when used as a comparative tool between surfaces known to be well cured and test areas, as long as sufficient repetitions are conducted.

Surface water absorption

Surface water absorption may be a potentially useful tool for assessing the quality of concrete curing (Figure 5.7; Poole 2006). The method appears to distinguish among different curing treatments. The most serious difficulty is that the method is sensitive to the moisture content of the concrete making it difficult to use in the field, but it can be used with cores that have been dried.

Abrasion resistance

Abrasion resistance may be one of the properties significantly affected by curing and may provide a useful means of field assessment (White and Husbands 1990). Laboratory testing by Poole (2006) showed that the

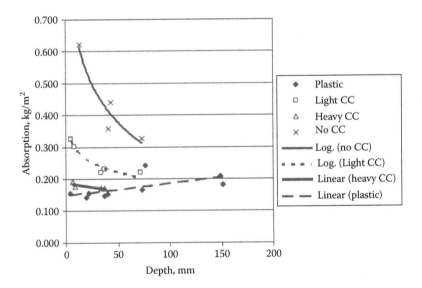

Figure 5.7 Surface water absorption versus depth for concrete cured with different methods. (From Poole, T. S., "Curing Portland Cement Concrete Pavements, Volume II," FHWA-HRT-05-038, Federal Highway Administration, McLean, VA, 2006.)

mortar fraction of poorly cured specimens had poor abrasion resistance compared with well-cured samples.

Chemically combined water

Measuring the amount of chemically combined water using thermogravimetric approaches can also be used to assess quality of curing. Mass loss between 110°C and 550°C will indicate the amount of water chemically bound in hydrated cement paste by hydration, but can also only be determined on cores, and requires a reasonably sophisticated laboratory. Each mixture will have to be calibrated for the relative proportions of paste, cementitious materials, and water.

Resistivity

McCarter et al. (1995) used a multielectrode resistance probe to investigate moisture movement within the *cover zone* of concrete. The test specimens were concrete slabs in which the probe set was attached to a bar 40 mm below the surface. The samples were cured using plastic sheeting for 7 days followed by 16 weeks drying in the laboratory before testing started. A head of water of about 200 mm was then applied to the top surface. The device indicated when external water reached the depth of probe. Some work would be needed to adapt the method to monitor curing but the technology appears feasible.

GUIDE SPECIFICATION

A generic document that seeks to act as a global or even national guide specification may not be ideal, because there are large variations in local materials, practices, and climate, let alone structural form. However, the basic guidelines that concrete should be kept wet and warm until it reaches some minimum level of performance can be used as the backdrop for any specification language.

Points that should be considered in a specification include the following.

Moisture control

- Materials
  - Water—Water should be free of materials that can stain the concrete surface.
  - Absorbent materials—AASHTO M182.
  - Sheet materials—ASTM C171, EN 14754, or local standards.
  - Curing (liquid membrane forming) compounds—ASTM C309 or C1315, or local standards.

- Evaporation retarders—There are no consensus standards covering these materials at present. Reference may be made to known reputable manufacturers.
- Internal curing materials—No standards exist for these products. ESCSI has published some guidance for lightweight aggregates. Superabsorbent polymers (SAPs) should be selected based on input from reputable manufacturers.
- Execution of initial curing (between placement and final finishing and when rapid drying is likely) can be by means of:
  - Fogging—The aim is to raise the relative humidity above the concrete surface, thus water should not collect on the concrete surface. The whole area to be protected must be adequately covered.
  - Using evaporation retarders—May be applied more than once if finishing is delayed.
- Execution of final curing (after bleeding has ended and final finishing) can be by means of:
  - Keeping the forms in place—Care must be taken to avoid rapid cooling or drying when forms are removed.
  - Fogging—The whole area to be protected must be adequately covered. Fogging must be continuous until the curing period is completed.
  - Sprinkling—The surface of the concrete must be sufficiently hard to prevent marring. Water temperature should not be more than 10°C cooler than the concrete. Ponding must be continuous until the curing period is completed. Runoff water must be controlled to avoid contaminating watercourses.
  - Ponding—The surface of the concrete must be sufficiently hard to prevent marring. Water temperature should not be more than 10°C cooler than the concrete. Ponding must be continuous until the curing period is completed. Runoff water must be controlled to avoid contaminating watercourses.
  - Using absorbent materials—The surface of the concrete must be sufficiently hard to prevent marring. Absorbent materials must be kept continuously moist until the curing period is completed. Absorbent materials should not be used without sheet materials to limit evaporation.
  - Using sheet materials—Sheet materials must be continuous either by taping or large overlaps. Sheets must extend beyond the end of the concrete and be secured on the edges to prevent acting as wind tunnels.
  - Using liquid membrane forming curing compounds—Selection of material is based on required efficiency, type of application, surface orientation, visibility, and whether other layers will be bonded to the concrete. All surfaces, including inside saw cuts should be treated. To be applied after bleeding and finishing is complete. It

is preferable to apply in two perpendicular coats to ensure more uniform coverage. To be protected from traffic and rain wash-off.
- Internal curing
  - Sufficient internal curing material is required in the mixture to provide the water required, but no more.
  - Materials must be saturated at time of batching.
  - Materials must be uniformly mixed.
  - Surface curing is still required.
- Application rates for curing compounds
  - Coverage rates to be high enough to ensure a continuous membrane, taking into account roughness of the surface and potential wind losses.
  - Two coats may be required to achieve sufficient covering.
  - Confirmation may be by means of referencing a color standard such as a white sheet of paper or the engineer's hardhat, or use of a reflectometer. Alternatively average coverage rates calculated by measuring area covered and volume of product used. The third alternative is to assess some performance metric such as rebound hammer or resistivity. Such approaches will have to be calibrated against each mixture used.
- Duration
  - This time is less critical in surfaces treated with curing compounds unless they are to be removed.
  - A number of approaches are available:
    - A fixed time.
    - A time based on system hydration rate and environment, typically between 3 and 21 days.
    - A fixed percentage of strength, typically 50% or 70%.
  - A required performance metric such as rebound hardness or resistivity.

Temperature control

- Cold weather precautions should be based on:
  - Protecting the concrete from freezing until sufficient maturity is achieved, thus keeping the system uniformly warm (and the exhaust fumes from heaters) and avoiding free water on the surface when freezing does occur.
  - Avoiding thermal gradients by applying too much heat at the surface.
- Hot weather precautions should be based on:
  - Avoiding rapid drying.

- Avoiding excessively high internal temperatures in order to improve final microstructure and avoid potential delayed ettringite formation (DEF).
- Avoiding excessive gradients when spraying with cold water.

General requirements

A guide specification should also provide checklists for the architect/engineer and specifier to ensure that all of the requirements are addressed. ACI 308 Guide Specification contains such guidance.

Information must be provided on how conformance with the specification will be assessed, that is, what test methods and measurement approaches will be used and at what frequency.

A specification should also provide guidance on the consequences of the failure to conduct the work as specified. Such actions may include the following:

- Reapplication of the curing method or an extension of the curing period.
- Repair of any cracking, up to and including removal and replacement of elements.
- Provision of surface sealers to make up for the loss of potential durability. The difficulty with this approach is that such materials have to be periodically replaced.
- Reduction in payments, based either on the value of the curing or up to the loss in lifetime of the structure, if it can be calculated.

PAYMENT

Payment for curing is a debated issue. Some specifications require that curing is conducted and no line item is provided for it. The disadvantage is that the contractor is not motivated to do the curing, and there is little action that can be taken if curing is not provided.

The only consequence of failure may be that the whole concrete element or structure is rejected, but in reality this is unlikely to be followed through. This approach is used in mass concrete elements and some slabs on grade where the risk of cracking is transferred entirely onto the contractor, forcing them to pay attention to preventive actions to control temperature differentials.

Some contracts provide a line item based on a surface area of concrete to be treated. The contractor will often put a low number in this item to keep the bid low and improve the probability of winning the contract. Again, there is little motivation to do the work and the punishment, even if no payment is made, for the item is normally small in terms of the whole contract value. Unless the contract explicitly provides a metric to assess whether the curing has been carried out satisfactorily, there may be disputes about whether payment is indeed due. This is especially true if the curing is in the form of spraying or flooding unless an inspector is permanently on site to monitor activities.

Other specifications may be based on the amount of curing compound that is used. This is likely the most practical approach to ensuring some moisture control. Care must be taken to ensure that the volume of product paid for is actually delivered, and used uniformly and in a proper manner.

Another alternative is to tie performance of the in-place concrete with the curing payment as discussed earlier. Problems with this approach include variability in performance test methods and potential for errors in batching also affecting in situ performance. Considerably more research is needed to prove this approach before it can be applied in live contracts.

REFERENCES

American Association of State Highway and Transportation Officials (AASHTO) M182, 2009, Burlap Cloth Made from Jute or Kenaf and Cotton Mats, M182-05.

American Concrete Institute (ACI) Committee 207, 1996, Mass Concrete, American Concrete Institute, Farmington Hills, MI.

American Concrete Institute (ACI) Committee 228, 1995, In-Place Methods for Determination of Strength of Concrete, American Concrete Institute, Farmington Hills, MI.

American Concrete Institute (ACI) Committee 305, 1999, Hot-Weather Concreting, ACI 305R-99, American Concrete Institute, Farmington Hills, MI.

American Concrete Institute (ACI) Committee 306, 1997, Cold-Weather Concreting, ACI 306R- 88, reapproved 1997, American Concrete Institute, Farmington Hills, MI.

American Concrete Institute (ACI) Committee 308, 2001, Guide to Curing Concrete, ACI 308R-01, American Concrete Institute, Farmington Hills, MI.

American Concrete Institute (ACI) Committee 308, 2011, Specification for Curing Concrete ACI 308.1-11, American Concrete Institute, Farmington Hills, MI.

ASTM C171-07, Standard Specification for Sheet Materials for Curing Concrete.

ASTM C309-11, Specification for Liquid Membrane-Forming Compounds for Curing Concrete.

ASTM C330/C330M-09, Standard Specification for Lightweight Aggregates for Structural Concrete.

ASTM C805/805M-08, Standard Test Method for Rebound Number of Hardened Concrete.

ASTM C1074, Standard Practice for Estimating Concrete Strength by the Maturity Method.

ASTM C1315-11, Standard Specification for Liquid Membrane-Forming Compounds Having Special Properties for Curing and Sealing Concrete.

ASTM E1347 11, Standard Test Method for Color and Color-Difference Measurement by Tristimulus Colorimetry.

Canadian Standards Association (CSA) A23, 2000, Concrete Materials and Methods of Construction/Methods of Test for Concrete, Canadian Standards Association.

Carrier, R. E., and Cady, P. D., 1970, "Evaluating Effectiveness of Concrete Curing Compounds," ASTM *Journal of Materials*, vol. 5, no. 2, pp. 294–302.

European Standard (EN) 206, Concrete—Specification, Performance, Production and Conformity.

European Standard (EN) 13670, Execution of Concrete Structures.

European Standard (EN) 14754, Curing Compounds—Test Methods—Part 1: Determination of Water Retention Efficiency of Common Curing Compounds.

Expanded Shale, Clay, and Slate Institute (ESCSI), 2012, "ESCSI Guide Specifications for Internally Cured Concrete." http://www.escsi.org/uploadedFiles/Technical_Docs/Internal_Curing/4001.1%20IC%20Guide%20Specification.pdf.

Kosmatka, S. H., and Wilson, M. L., 2011, *Design and Control of Concrete Mixtures*, 15th ed., EB001, Portland Cement Association, Skokie, IL.

Kropp, J., and Hilsdorf, H. K., 1995, *Performance Criteria for Concrete Durability*, E&FN Spon, London.

McCarter, W. J., Emerson, M., and Ezirim, H., 1995, "Properties of Concrete in the Cover Zone: Developments in Monitoring Techniques," *Magazine of Concrete Research*, vol. 47, no. 172, pp. 243–251.

Meeks, K. W., and Carino, N. J., 1999, "Curing of High-Performance Concrete: Report of the State-of-the-Art," NISTIR 6295, National Institute of Standards and Technology, Gaithersburg, MD.

National Ready Mixed Concrete Association, 1960, "Plastic Cracking of Concrete," Engineering Information, NRMCA, Silver Spring, MD.

Perrie, B. D., 1994, "The Testing of Curing Compounds for Concrete," master's dissertation, University of the Witwatersrand, Johannesburg, South Africa.

Poole, T. S., 2001, "Methods for Measuring Application Rate of Liquid Membrane-Forming Curing Compounds on Concrete Pavements," *Proceedings of the 7th International Conference on Concrete Pavements, International Society for Concrete Pavements*, pp. 351–360.

Poole, T. S., 2006, "Curing Portland Cement Concrete Pavements, Volume II," FHWA-HRT-05-038, Federal Highway Administration, McLean, VA.

Senbetta, E., 1994, "Curing and Curing Materials," in *Significance of Tests and Properties of Concrete and Concrete-Making Materials*, P. Klieger and J. Lamond, eds., ASTM STP 169c, pp. 478–483.

Shariat, S. M. S., and Pant, P. D., 1984, "Curing and Moisture Loss of Grooved Concrete Surfaces," *Transportation Research Record*, 986, pp. 4–7.

Unified Facilities Guide Specifications (UFGS), 2012, 03 39 00 Division 3, Concrete Curing.

Vandenbossche, J. M., 1999, "A Review of the Curing Compounds and Application Techniques Used by the Minnesota Department of Transportation for Concrete Pavements, Minnesota Department of Transportation," St. Paul, MN.

White, C. L., and Husbands, T. B., 1990, "Effectiveness of Membrane-Forming Curing Compounds for Curing Concrete," Miscellaneous Paper SL-90-1, U.S. Department of Army, Waterways Experiment Station (now Engineer Research and Development Center), Vicksburg, MS.

Chapter 6

Real-world experience

INTRODUCTION

This chapter is a review of reports that describe work conducted in the field around the world to evaluate or compare performance of different curing systems. The papers have been grouped based on similar concerns such as hot climate, cold climate, high-performance concrete, and other performance parameters. One section also addresses work conducted to help development of specifications. Papers are discussed in chronological order.

Discussion is also provided for each paper that puts the findings into the context of this publication or qualifies some of the findings based on consensus of the body of literature.

HOT CLIMATE

Alsayed and Amjad (1994)

Alsayed and Amjad (1994) reported work conducted to assess the effect of desert-type curing on concrete performance. They formed slabs and 150×300 mm cylinders, exposed them to four curing regimes, and tested them for compressive strength, modulus of elasticity, porosity, water absorption, and shrinkage behavior.

Samples were covered with burlap in the molds for 24 hours before demolding and "cured" for 7 days. One set was sprinkled twice a day, another was covered with burlap and sprinkled with water twice a day, a third set was covered in plastic, and the fourth was exposed to the weather. The samples were then exposed for a year to a desert climate.

Initial shrinkage readings were taken at the end of the curing period. Other tests were conducted at the end of the exposure period. Strength was evaluated using 150 mm cores extracted from the slabs. Twenty-eight-day strengths were also reported, presumably using formed cylinders. The data are summarized in Table 6.1.

Table 6.1 Test Data for Varying Curing Techniques

	fc, 28 days (MPa)	fc, 360 days (MPa)	E (GPa)	Porosity (%)	Absorption (%)
Sprinkled	36.3	39.8	31.1	11.40	5.64
Burlap	36.3	39.7	29.6	11.95	6.10
Plastic	45.6	32.5	27.5	14.04	6.51
None	38.5	27.9	26.2	14.82	7.28

Source: Data from Alsayed, S. H., and Amjad, M. A., *Cement and Concrete Research*, 24, 7, 1390–1398, 1994.

The mixture used had a target 28-day strength of 35 MPa, and contained ordinary portland cement (Type I), 20 mm coarse aggregate, washed desert sand, and a water to cementitious materials ratio (w/cm) of 0.45. Also samples reportedly contained reinforcing steel.

The authors explained that the difference in strength between the plastic-wrapped set and the two added-water sets was due to differences in heat of hydration, with the plastic sheet acting as an insulator keeping heat in. They also considered that the plastic wrapped samples had insufficient water to continue hydrating for the year of exposure as evidenced by the drop in strength of both the uncured and plastic wrapped samples. The small difference between the sprinkled and burlap set indicated that there was little benefit associated with the use of the burlap.

It was acknowledged that continuous wetting may have yielded different performance, but there was concern that this may not be possible in arid locations where water is in limited supply.

The drop in strength was a matter of concern and suggestions were made that the design codes should be modified to accommodate arid climates and the potential reduction in strength over time. It was noted that the modulus elasticity data could be predicted from the compressive strengths, regardless of the curing provided.

Porosity and permeability were markedly better in the water-added samples than in the other two sets including the plastic wrapped. It was argued that the severe environment these samples were exposed to explained the poor performance of the plastic wrapped set, and so in extrapolation the poor performance of many local structures built in accordance with specifications from countries with temperate climates.

Moisture movement was assessed by comparing shrinkage rates. Initial shrinkages were rapid in all cases (Figure 6.1). The initial shrinkage of the plastic wrapped samples was less than the added-water samples, and ultimate shrinkage of all of the sets was considered to be similar, albeit at different rates.

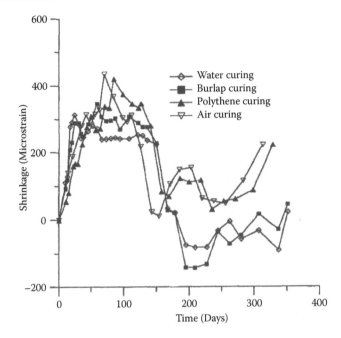

Figure 6.1 A plot of shrinkage of field-exposed samples over time. (From Alsayed, S. H., and Amjad, M. A., *Cement and Concrete Research*, 24, 7, 1390–1398, 1994. With permission.)

Discussion

The data and some of the findings presented in this work warrant some discussion. As noted in Chapter 3, the magnitude of the effect of curing on strength of a sample will be strongly affected by the size of the sample, with larger affects to be observed in smaller samples. If different sample sizes were used, for example cylinders and cores from larger slabs, at 28 and 360 days, comparison of strengths may not be appropriate. The high early strengths of the dryer sets may not be unexpected, because a sample tested dry will normally exhibit higher strengths than similar, wetted samples. The samples tested at one year are likely to be in a similar state of saturation; thus comparison between them would more accurately reflect the extent of hydration, greatest in the early wet samples and least in the samples allowed to dry out sooner.

The findings regarding the plastic sheeting indicate that the plastic had little moisture-related benefit. This is supported by the poor absorption results exhibited by the samples wrapped in plastic. It calls into question whether the plastic was sufficiently sealed around the samples to prevent moisture loss, because at a w/cm 0.45 self-desiccation would not be expected, and other work has clearly shown the benefits of effective wrapping. The alternative is that the period over which it was applied was too short.

The shrinkage data are also interesting. The paper does not provide details of the weather over the exposure period, but the large expansions that started at about 120 days are greater than the initial shrinkage. This behavior is unexpected and may only be explained by changes in temperature and moisture state, likely due to changes in season.

The findings are useful though in that they demonstrate the importance of keeping in mind all of the factors that may affect the performance of a given mixture, and how that performance is measured. They also strongly support the contention that curing is critically important for structures exposed to severe environments, particularly from the point of view of potential durability. In such environments it may be advisable to avoid self-desiccating mixtures. Added water approaches may be desirable, but difficult to achieve in arid areas, making it even more important to pay particular attention to effective prevention of premature drying, whether by wrapping in plastic or the use of curing compounds.

Hoppe et al. (1994)

The objective of Hoppe et al.'s study (1994) was to investigate the reliability of South African–developed durability index tests when they are used to assess the effectiveness of site curing.

The work comprised extracting cores from a New Jersey barrier built in Cape Town (Mediterranean climate), one side of which had not been cured and the other had curing compound applied.

The concrete delivered was a standard mixture supplied by a ready-mix operation. The mixture comprised 310 kg/m³ with 15% fly ash (similar to Class F) and a w/cm of 0.54. Forms were removed 24 hours after placement. Curing compound was applied on one face, except the application was anywhere up to 6 hours after the forms were removed because surface defects had to be repaired. The water-based product was applied by brush because strong winds made spraying impractical.

Sampling and testing was carried out twice during construction: once in the summer and the other 3 months later. The weather during the first sampling was hot, dry, and windy while during the second sampling it was moderate, with occasional wind and light rain. Cube samples were made from the as-delivered concrete for lab testing, and cores were extracted from the barrier after 28 days.

All the experimental data show that concrete that was exposed was more permeable than concrete that had been kept continuously wet. The curing compound was marginally effective in the summer placed concrete, and surprisingly, led to worse performance in the moderate weather placement compared to the exposed surfaces. It is believed that the late application of the compound allowed the concrete to dry significantly before it was treated. Later rains in the moderate weather thus enhanced hydration, while the treated surfaces were kept dry.

Reasons for the relatively poor effectiveness of the curing compound are likely:

- Delays between stripping and application of the curing compound likely caused significant drying, particularly during summer.
- Poor proportioning and mixing of the curing compound may have reduced its effectiveness.
- Application rates of the curing compound may not have been sufficient.
- Form release compounds may have impaired the performance of the curing compound.
- The curing compound developed for European conditions may not have been adequate for harsh conditions in South Africa.

It was recommended that performance tests such as those used in this work should be used to monitor curing efficacy on concrete in place.

Discussion

Improper use of curing compounds, such as delayed application, appears to have led to limited benefit from its use, particularly when the weather is moderate. A lack of effective curing led to marked loss in potential durability of the system.

Al-Gahtani (2010)

Al-Gahtani (2010) conducted a study to evaluate the effect of curing methods on the properties of concretes in order to improve specifications. The objective of curing is to keep concrete wet enough to promote cement and supplementary cementitious material (SCM) hydration. This is difficult to achieve in hot, dry climates. The effect of curing is more important for permeability than it is for strength.

Concrete mixtures contained plain cement, very fine fly ash (10%), silica fume (0%), and fly ash (30%). The binder content was 370 kg/m^3 with a line feed w/cm of 0.45. Curing was provided in the form of

- Wet burlap
- Acrylic curing compound
- Water-based curing compound

Samples were coated with epoxy after 24 hours on all but one face to ensure that measured effects were localized to one face.

The following properties were monitored:

- Plastic shrinkage for 24 hours
- Drying shrinkage
- Pulse velocity
- Compressive strength at 3, 7, 14, 28, and 90 days

Plastic shrinkage varied with each binder type (Figure 6.2), but the effect of curing was similar for all the binders tested. The general trend was that the most shrinkage was observed in the exposed samples and the least was with the use of acrylic curing compound.

Discussion

Providing protection with curing compounds appeared to be more effective than use of burlap in terms of shrinkage, strength gain, and pulse velocity for all the cementitious systems tested.

Likewise the drying shrinkage was most in the burlap samples and least under acrylic curing compound, although the differences were relatively small (Figure 6.3). Strength development and ultrasonic pulse velocity (Figure 6.4) was also greatest in the acrylic cured samples and least in the burlap.

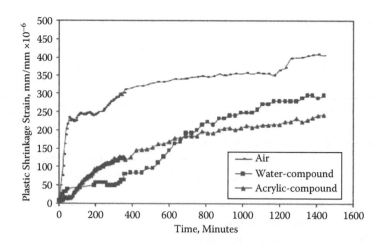

Figure 6.2 A typical plot of plastic shrinkage for the different curing techniques. (From Al-Gahtani, A. S., *Construction and Building Materials*, 24, 308–314, 2010. With permission.)

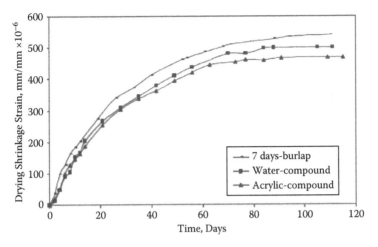

Figure 6.3 A typical plot of drying shrinkage for the different curing techniques. (From Al-Gahtani, A. S., *Construction and Building Materials*, 24, 308–314, 2010. With permission.)

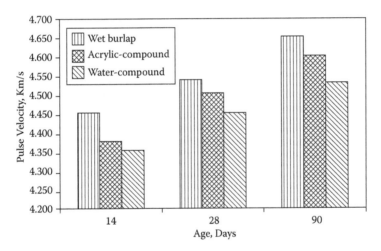

Figure 6.4 Typical variation of pulse velocity with time and curing technique. (From Al-Gahtani, A. S., *Construction and Building Materials*, 24, 308–314, 2010. With permission.)

COLD CLIMATE

Boyd and Hooton (2007)

The Ministry of Transportation of Ontario (MTO) has limited dosages of SCM contents for bridge decks and barriers because of concerns about scaling resistance. Higher slag and fly ash contents have been shown to perform satisfactorily in the field despite laboratory data indicating otherwise.

Boyd and Hooton (2007) investigated the effect of SCMs, dosage, and curing on scaling performance.

Large-scale slabs were exposed to 12 years of natural freeze–thaw cycling along with regular salting and traffic. The cementitious materials used were a CSA Type 10 normal portland cement, slag (25% to 50%), and a Class C fly ash (10% to 15%) in six combinations, including one ternary mix and a plain cement control. Air was entrained using a modern tall-oil product. Mixture w/cm was 0.42 for a design strength of 32 MPa, and a target air content of 5% to 8%.

Compressive cylinders 100×200 mm were used to monitor compressive strength of the concrete, half of which were cured normally and tested at 7 and 28 days. The other half were buried next to slabs and tested at 128 days. A $250 \times 350 \times 75$ mm slab was prepared using each mix and moist cured to 28 days for pull-off testing.

Other slabs were cast in situ: one set in a heavy truck traffic zone, others in a light traffic zone. Thermocouples were used to monitor freeze–thaw cycles. Each slab was divided into sections with different curing regimes:

- 4 days with a burlap and plastic
- Curing compound

Other variables were the finishing: after bleed water disappears and prematurely. Laboratory scaling samples were also prepared using standard curing methods along with the aforementioned curing methods, and were tested in three different laboratories. Similar samples were exposed for 128 days outdoors before testing.

Twenty-eight-day strengths ranged between 35 and 45 MPa, increasing to 42 to 57 MPa at 128 days. Laboratory scaling tests indicated that increasing

Table 6.2 Mass Loss Data for Varying Binder Type and Curing Technique under Salt Scaling Testing (kg/m²)

	Burlap and Plastic		Curing Compound	
	Early Finish	Normal Finish	Early Finish	Normal Finish
50% Slag	1.28	0.81	1.12	0.68
35% Slag	0.15	0.08	0.08	0.12
25% Slag	0.12	0.05	0.08	0.14
25% Slag + 10% Fly Ash	0.07	0.44	0.11	0.15
15% Fly Ash	0.41	0.26	0.05	0.07
100% OPC	0.05	0.06	0.07	0.08

Source: Boyd, A. J., and Hooton, R. D., Journal of Materials in Civil Engineering, 19, 820–825, 2007. With permission.

SCM content decreased scaling resistance (Table 6.2). There was considerable variability between laboratories. Slabs exposed for 128 days were notably better. No trend was found to correlate finishing timing with scaling resistance. A clear trend was observed with changing curing of the laboratory-tested samples. Resistance to scaling was better in the samples treated with curing compound than in those covered with burlap and plastic.

For the slabs exposed in the field for 10 years and about 600 freeze–thaw cycles, only the 50% slag mixture showed any visible (and minor) scaling, limited to the sections not treated with curing compound.

A steel disk was bonded to the surface of the bond-test samples, and the force required to pull it off was recorded. Addition of fly ash appeared to increase surface strength, while slag reduced it. Variation in finishing had little effect while curing had more influence. The best performance was once more obtained from the samples cured with curing compound.

It was noted that placing a slab well before the first freezing event would be expected to perform better than a slab placed closer to winter.

Discussion

The primary purpose of the work was to correlate test methods and practices with field performance but did demonstrate the benefit of curing compound in increasing scaling resistance.

Bouzoubaâ et al. (2011)

Bouzoubaâ et al. (2011) conducted a rigorous field and laboratory study of salt scaling of sidewalk mixtures including the effects of curing regimes. Reference was also made to previous work conducted by the same team (Bouzoubaâ et al. 2008).

Sidewalks and laboratory slabs were constructed during the Canadian fall to observe the field performance of mixtures prepared late in the season. Temperatures fell to below freezing overnight during the time the concrete was placed and cured. Three mixtures were used containing plain portland cement, 25% fly ash, and a ternary cement containing fly ash and silica fume. The mixtures had a target minimum w/cm of 0.45 and air content between 5% and 8%. The reported fresh properties of the mixtures were as uniform between mixtures as would be expected. Curing comprised use of a curing compound, or 2 days covered with wet burlap and plastic.

Cores were extracted from the laboratory slabs at 2 days, and then tested for strength (Table 6.3) and salt scaling. The slabs were left outdoors and further cores taken at 28 days. Standard cylinders were also prepared and tested in compression. The sidewalk sections were visually inspected for distress for 6 years.

Table 6.3 Compressive Strength Data, MPa for Field Cores

	3 Days			28 Days		
	Cylinders	Curing Compound	Wet Burlap	Cylinders	Curing Compound	Wet Burlap
Plain cement	22.8	22.9	23.9	38.1	36.4	34.8
Fly ash	20.4	15.4	17.1	31.2	25.9	27.6
Ternary	19.8	15.5	16.9	40.0	28.3	26.9

Source: Data from Bouzoubaâ, N., Bilodeau, A., Fournier, B., Hooton, R. D., Gagné, R., and Jolin, M., *Canadian Journal of Civil Engineering*, 38, 373–382, 2011. With permission.

The strength data from the cores show no significant effect of the type of curing. The strength development of the fly ash mixture was slower than the other mixtures, particularly in the core samples. This is believed to be due to the low temperatures experienced by the slabs having a more marked effect on the fly ash than the other cementitious systems. This is supported by earlier work that showed little difference between cylinders and cores from slabs prepared in warmer weather (Bouzoubaâ et al. 2008).

The cores kept under burlap generally did not perform as well in scaling testing (Figure 6.5) as those sprayed with curing compound, particularly in the SCM mixtures. This is believed to be related to the rate of drying and associated microcracking when burlap is removed.

The plain cement mixtures performed well in the field (Table 6.4), with reducing scaling resistance in the SCM mixtures. The sections sprayed

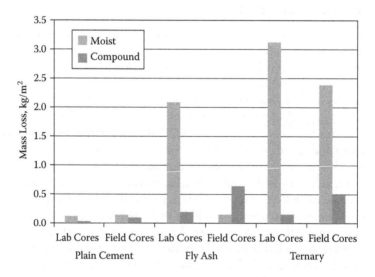

Figure 6.5 Mass loss of cores tested under salt scaling. (Redrawn from Bouzoubaâ, N., Bilodeau, A., Fournier, B., Hooton, R. D., Gagné, R., and Jolin, M., *Canadian Journal of Civil Engineering*, 35, 11, 1261–1275, 2008. With permission.)

Table 6.4 Visual Rating of Sidewalks

	Curing Compound	Wet Burlap
Plain cement	1	1–2
Fly ash	2–3	3
Ternary	3	3–4

Note: Based on ASTM C672.

Source: Data from Bouzoubaâ, N., Bilodeau, A., Fournier, B., Hooton, R. D., Gagné, R., and Jolin, M., Canadian Journal of Civil Engineering, 38, 373–382, 2011. With permission.

with curing compound were marginally better than those covered with wet burlap. In comparing results with those from the previous work, the slabs placed in spring performed better than those placed in the fall, likely due to greater maturity when freezing weather started.

The conclusions reached included that salt scaling resistance was best with the use of a curing compound and in mixtures placed in the spring.

Discussion

It is noted that the fly ash and ternary mixtures did not perform well, and were more sensitive to maturity and curing than the plain cement. This is despite the literature generally indicating that potential durability is improved with the use of SCMs.

It is likely these observations are a combination of the following factors:

- Scaling resistance is likely strongly influenced by system degree of hydration, with large benefits being obtained from small increments of hydration, especially in more mature systems. Such increments are unlikely to have a noticeable effect on strength.
- Systems containing SCMs are known to hydrate more slowly. This means that 2 days of burlap is unlikely to be sufficient to achieve the levels of hydration in the surface layer needed to provide scaling resistance.
- Curing compounds are likely to provide equivalent curing somewhat better than 2 days under burlap, both in terms of the extent of hydration at the surface and the risk of cracking with rapid drying when the burlap is removed.

HIGH-PERFORMANCE CONCRETE

Huo and Wong (2006)

Unexpected shrinkage and temperature cracks can reduce the strength, durability, and serviceability of concrete. Huo and Wong (2006) conducted a program to investigate the early age behavior of high-performance concrete (HPC) specimens subjected to different curing methods. Properties measured included shrinkage, temperature change and evaporation rate.

The mixture used was sampled from a mixture used in a bridge deck in Tennessee and contained 9% silica fume with a w/cm of 0.333. The samples made were $600 \times 1200 \times 140$ mm slabs and 150×300 mm cylinders. Curing regimes applied were

- Wet burlap blankets
- Cotton mats
- Curing compound (two applications)
- Polyethylene blankets

Curing was applied within 30 minutes of casting and continued for 3 or 7 days.

Temperatures were monitored using thermocouples in both types of samples for 30 hours. Strains were monitored using Demec discs attached to surfaces of the cylinders and slabs after the curing material was removed. Compressive strengths were measured at 3, 7, 14, 28, and 56 days, and modulus of elasticity of the HPC was measured at 28 days using conventionally cured cylinders. Compressive strength at 28 days was 65.7 MPa and the modulus of elasticity was 35.6 GPa.

The highest temperatures were recorded in the samples under plastic sheeting and curing compound (Figure 6.6). It was speculated that this was because of evaporative cooling of the burlap and cotton mats. Cylinders

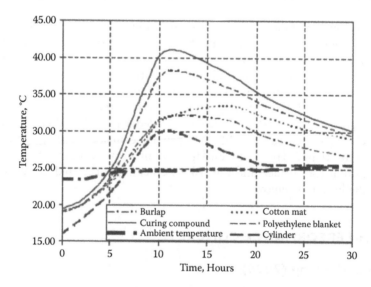

Figure 6.6 Plot of concrete temperatures for varying protection systems. (From Huo, X. S., and Wong, L. U., *Construction and Building Materials*, 20, 1049–1056, 2006. With permission.)

exhibited lower temperatures than the slabs, likely due to the greater surface to volume ratio.

Shrinkage started earlier in the samples treated with curing compound
than in the other samples (Figure 6.7). This is likely because the curing
compound did allow some moisture loss almost as soon as drying started.
The other samples started shrinking when the materials were removed. The
lowest shrinkage was observed in the samples under plastic sheeting. It is
notable that the ranking of shrinkage was established within a few weeks
of casting, and that the shapes of the curves are all similar, indicating the
sensitivity of the test to the ambient environment and the importance of
early protection.

Moisture losses were recorded by monitoring mass of cylinder samples (Figure 6.8). The rate of moisture loss was similar for all the curing
approaches (within the precision of the method used) and strongly influenced by the age at which drying started. The rate of drying decreased over
time, particularly for the first 5 days.

A cost comparison was also carried out and it was found that the curing
compound was about 20% of the cost of the other approaches.

Discussion

This work was interesting in showing the relative drying and shrinkage
effects of the common curing approaches. Again the importance of early
protection for about 7 days was highlighted.

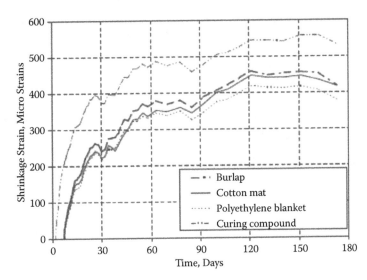

Figure 6.7 Shrinkage data. (From Huo, X. S., and Wong, L. U., *Construction and Building
Materials*, 20, 1049–1056, 2006. With permission.)

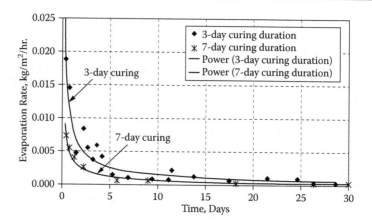

Figure 6.8 Evaporation rates. (From Huo, X. S., and Wong, L. U., *Construction and Building Materials*, 20, 1049–1056, 2006. With permission.)

Poursaee and Hansson (2010)

Concrete shrinkage, chemical reactions, and thermal stresses are factors that may cause cracks in concrete and can be controlled to a large extent by appropriate curing, especially HPC.

HPC is at increased risk of self-desiccation because of the low w/cm, which can result in autogenous shrinkage. Some recommend that HPC should be cured for a longer time to overcome this tendency. This extended curing requirement can increase costs and cause delays. The work by Poursaee and Hansson (2010) sought to investigate whether increasing curing from 3 to 7 days was beneficial or required.

The work comprised monitoring internal strain, moisture content, and temperature of samples for 100 weeks.

Three concrete mixtures were prepared. One mixture, considered a regular concrete control, had a w/cm of 0.43 and contained no SCMs. The other two test mixtures, considered high performance concrete (HPC) had a w/cm of 0.35 and contained either 25% fly ash or slag.

Prisms $500 \times 100 \times 100$ were prepared from the mixtures. Sensors were cast inside the prisms at depths of 25 mm and 50 mm from the surface. Samples were subjected to wet curing periods of 2, 3, or 7 days in wet burlap and a plastic sheet. Samples were kept in a tent from the time of casting for 14 days before they were exposed to the Canadian weather for 100 weeks.

The sensors monitored strain, temperature, and moisture content. The strain gauges used had a small gauge length meaning that localized deformations in the paste were recorded. Moisture sensors were based on an innovative approach of measuring electrical resistivity of small wooden bars that was known to vary with changing moisture content, and calibrated

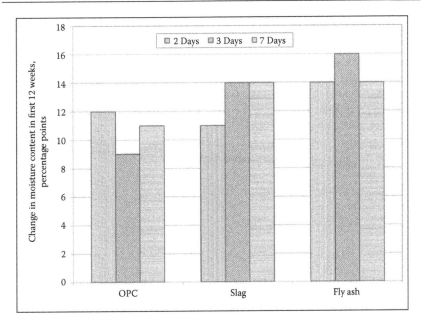

Figure 6.9 Internal moisture content at 50 mm from the surface of the prisms: OPC, slag, and fly ash. (Data from Poursaee, A., and Hansson, C. M., *Proceedings of the Institution of Civil Engineers Construction Materials*, 163, 4, 223–230, 2010.)

accordingly. An additional temperature sensor was placed externally to the samples to record air temperature.

The data indicate an approximate 2°C to 5°C temperature difference between interior and exterior sensors. During cyclic changes the interior temperature lagged the exterior by only 20 minutes regardless of the curing provided.

The greatest difference in moisture contents over time was about 15%. Moisture movement seemed to slow over time, likely due to increased concrete maturity. The control mixture exhibited a lower moisture content than the test mixtures. There was no significant difference in moisture movement with different curing times (Figure 6.9).

In all prisms, the strain increased over time, with lower strains observed in the control mixture and little difference between the slag and fly ash mixtures. The data show that there was no clear trend with respect to the effect of wet curing time on the internal strain.

Discussion

The findings in this paper are somewhat surprising and the reasons for the lack of trend with curing time cannot be explained from the information available.

Zhimin and Junzhe (2011)

Zhimin and Junzhe (2011) examined the effects of steam curing and binder type on sorptivity at different depths of concrete samples. Precast concrete elements are often heated with steam to accelerate strength development. However, steam curing is reported to negatively affect the quality of the cover layer in terms of cracking risk and increased sorptivity. Increased sorptivity may be considered an indicator of reduced durability because water transport plays a critical role in deterioration mechanisms of concrete.

Three mixtures were prepared with a w/c of 0.27:

- Ordinary portland cement (OPC)
- 20% fly ash, 10% slag
- 25% fly ash, 5% silica fume

Some samples were moist cured for 28 days while others were steam treated in the following cycle:

- Preheating and heating, 2 hours
- 60°C, 8 hours
- Cooling, 1 hour

Cores were cut from slabs and cut into 48 mm thick slices with top faces of the slices from the outer surface, and 10 mm and 30 mm below the surface.

Water sorption was conducted on the top faces of the samples up to 24 hours (Figure 6.10).

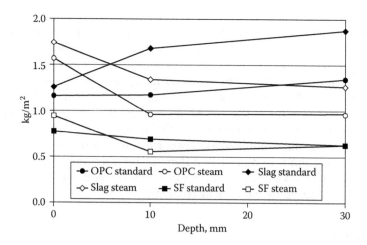

Figure 6.10 Water sorption data. (Data from Zhimin, H. E., and Junzhe, L. I. U., *Key Engineering Materials*, 477, 263–267, 2011.)

For all the mixtures subjected to steam curing the sorptivity decreases between the surface and the next layer; thereafter there is little change. The best performance under steam was exhibited by the silica fume mixture, followed by the OPC mixture, then the ternary mixture with fly ash and slag.

The performance of the standard cured samples is less uniform. The silica fume mixture improved with depth, the OPC mixture did not change significantly, and the ternary mixture got worse. The ranking of the mixtures was the same as for the steam-cured samples.

The effect of the steam curing was to improve the system at depth in all mixtures but to make it worse than moist curing at the surface in all cases.

Discussion

Some of the findings are consistent with the literature, while some of the others are unexpected. The poor performance of the ternary mixture with respect to the OPC system is not consistent with reported data. It is possible that the chemistry of the cementitious materials meant that they were not well suited to be combined in the proportions selected.

On the other hand the negative effect of steam on the outer surface is in line with experience. Unpublished work conducted by the author demonstrated that steam curing had a negative effect at all the depths tested. This was explained as being due to accelerated hydration leading to a coarser and thus more permeable microstructure.

PERFORMANCE PARAMETERS

Petrou et al. (2001)

Petrou et al. (2001) investigated cracking of bridge decks. The South Carolina Department of Transportation adopted use of an HPC mix design containing both microsilica and fly ash with a w/cm of 0.37 for use in bridge decks in order to improve potential durability. However, a number of decks built with the mixture have exhibited problems with early age cracking. The aim of the reported work was to examine factors contributing to the cracking.

Nine bridges that had been constructed in the 1990s were evaluated. Most were multiple-span continuous steel girder structures. The decks were either conventional or HPC concrete.

Bridge 1 is a steel girder multiple-span continuous structure with a conventional deck. Significant full-width cracks appeared shortly after the span was opened to interstate traffic. The cracks ran perpendicularly across the structure and varied from a hairline to about 3 mm wide. The cracks were aligned with the barrier wall control joints. Cracks located over the piers were observed to extend through the deck slab.

Bridges 2 and 3 are steel girder multiple continuous-span structures with HPC decks. Both spans exhibited significant shrinkage cracking shortly after placement. It was noted that conditions were windy during some of the pours preventing use of curing mats. There is some full-width hairline cracking extending from control joints in the barrier walls.

Bridge 4 is a curved multiple-span continuous bridge on straight steel girders with a conventional deck. There are no control joints in the integrally cast barrier walls. Some full-width cracks were observed but they were smaller with a closer spacing.

Bridges 5 and 6 are short simple skew spans on steel girders with HPC decks. Both spans exhibit some shrinkage cracking and some cracks that appear to have occurred since the structures were first loaded. The cracks on Bridges 5 and 6 are perpendicular to the skew.

Bridge 7 is a steel girder four-span continuous bridge, and Bridge 8 is a nearby concrete girder three-span continuous structure. Both bridges have HPC decks. Both structures exhibit significant shrinkage cracking in isolated pour segments. It is considered likely that the cracking resulted from poor curing practices. There is some limited full-width cracking initiated at control joints in the barrier walls.

Bridge 9 is a steel girder three-span continuous bridge with an HPC deck. Very little cracking was observed on this bridge, likely resulting from more efficient curing practices.

A number of potential contributors to cracking risk were considered:

1. High evaporation rate—Records show that curing requirements were not adequately met on Bridges 2 and 3, nor, probably, on Bridges 7 and 8. The cracks were considered to be likely the result of poor curing based on winds preventing the use of curing mats and reported low core strengths. The cracking in Bridges 7 and 8 was only observed in isolated sections also suggesting poor protection.
2. Use of high-slump concrete—Data for Bridges 2 and 3 indicate that the concrete had a slump between 75 and 100 mm. This range is believed to decrease bond with the reinforcing bars, and is more vulnerable to the effects of machinery and traffic vibration. In general, little pattern cracking was observed on these decks; therefore slump is not likely to be significant in this case.
3. Excessive water—Concerns were raised that there was excessive water in the mixtures used in Bridges 2 and 3. This water was likely due to wash water being left in the concrete trucks and inadequate control of moisture content in fine aggregate. However, review of the documentation indicated that the amount of water actually used was less than that specified in the mix design.

4. Insufficient top reinforcement—The amount of top reinforcement was not directly investigated, but the spacing of cracking indicates that the reinforcement is not a contributing factor to the observed cracking.

In summary, significant early age cracking was reported for Bridges 2, 3, 7, and 8. No additional movement was observed after they were sealed during the construction period indicating that the cracks were not related to structural movements. The other bridges inspected exhibited very little or no early age cracking regardless of the mixture. There did not seem to be a relationship between cracking risk and mixture type.

All bridges inspected exhibited some degree of load-induced cracking in the form of full-width transverse cracks. In all cases, these cracks are spaced uniformly along the bridge span. Two initiators were observed: control joints in integrally cast barrier walls and at deck drains. Bridge 4 had no control joints in its integrally cast barrier walls and the load-induced cracking was more controlled in this case.

The authors have concluded that poor curing was a major contributor to the cracking in four of the decks investigated. In addition, careful design is required to balance structural deflections with mixture stiffness, as well as structural strength, permeability, and cracking risk.

Discussion

The discussion about the influence of slump on cracking risk needs to be approached carefully. Shrinkage is directly controlled by the paste–water content of a mixture. When slump was primarily controlled by water content in a mixture there was a relationship between slump and cracking, but in current technology where slump is controlled by the use of chemical admixtures, this relationship is no longer necessarily applicable.

Johnston and Surdahl (2007)

Johnston and Surdahl (2007) investigated the relationship of some factors affecting cracking in continuously reinforced concrete (CRC) pavement in South Dakota.

Two mixtures were used in the pavements with the critical differences shown in Table 6.5.

During construction temperatures ranged from 50°F to 91°F, wind speed ranged from 0 to 25 mph, and relative humidity ranged from 18% to 97%.

Crack frequencies were determined for each 500 ft section of pavement. Concrete temperature was shown to have the strongest correlation to cracking, along with a substantial reduction in the level of cracking when the concrete mixture was changed. In addition, the section in which curing compound residue was still visible 1 year after placement exhibited minimal cracking.

Table 6.5 Descriptions of Mixtures

	Mix 1	Mix 2
Cementitious (pcy)	600	575
Fly ash (%)	20	19
w/cm	0.405	0.395
Nominal aggregate size (inches)	1	1.5

Source: Johnston, D. P., and Surdahl, R. W., Transportation Research Record: Journal of the Transportation Research Board, Transportation Research Board of the National Academies, Washington, DC, No. 2020, 83–88, 2007. With permission.

Statistical modeling of the data also found that temperature difference between the air and concrete was the critical variable. This meant that wind velocity had a beneficial effect on cracking because it allowed more rapid equilibration between the air and concrete surface, despite the increased risk of plastic shrinkage cracking. This also highlighted the benefit of cooling the concrete mixture by means such as sprinkling the aggregate stockpile.

The w/cm was also found to have a counterintuitive effect with higher water content mixtures exhibiting less cracking, likely due to the lower water content mixtures being difficult to consolidate.

Two additional sections were constructed using mixture containing 622 lb/yd³ of cementitious material. A normal curing compound application rate was used on one section, and the rate was doubled on another. Results showed a marked improvement in cracking rates with the increased curing.

Based on these findings the following recommendations were made:

- Curing compound should be applied to the pavement within 30 minutes at a minimum rate of 1 gal/125 ft², with a second application at the same rate applied within another 30 minutes.
- The aggregate stockpiles should be wetted down at the conclusion of daily production when temperatures greater than 80°F are expected the following day.
- Cementitious content should be 622 lb/yd³ to ensure adequate water content for consolidation and sufficient strength gain rates.

Discussion

The findings support the contention that a number of factors may contribute to cracking risk. Although some factors may have a large effect, such as temperature, they are not easily controllable; therefore changing the albeit more minor factors that can be adjusted, such as enhanced curing, can still have a marked benefit.

Radlinski et al. (2008)

Radlinski et al. (2008) used maturity measurements to monitor curing. Maturity is a technique that allows prediction of the compressive strength of a concrete based on the product of time and temperature experienced by the mixture. It is primarily used as a tool to determine the time for opening to traffic or formwork removal.

Part of the aim of the work was to assess the validity of using the technique to monitor mixtures containing supplementary cementitious materials.

Laboratory concrete

Concrete mixtures were prepared using Type I portland cement, Class C fly ash (FA), and silica fume (SF). Four mixtures were prepared containing both SCMs at 20 or 30 fly ash and 5% or 7% silica fume. The w/cm ratio was fixed at 0.41 and the binder content was 231 kg/m^3.

Samples were subjected to five different curing/drying conditions:

- In water
- In air
- 7 days curing compound which was then brushed off
- 3 days under wet burlap and plastic sheet
- 7 days under wet burlap and plastic sheet

After curing, the samples in the last three sets were stored at 23°C and relative humidity (RH) 60%.

Concrete temperature was monitored using Type T thermocouples embedded in 102×203 mm cylinders. Cylinder compressive strengths were determined at 1, 3, 7, 28, and 90 days.

The maximum temperature of specimens covered with burlap was about 2°C higher than the maximum temperature of specimens in air or curing compound. This may be a benefit in cold weather. A 2°C drop was observed at the time air-exposed specimens were demolded and at the time the burlap samples were unwrapped. This is likely because of evaporative cooling when the moist surfaces were exposed.

The compressive strengths were compromised by decreasing quality of curing with the highest strengths observed in the samples stored in water. In addition, the strengths of mixtures containing 30% FA were more affected by poor curing than the 20% FA mixtures.

The prediction of compressive strength of the mixtures was based on the equivalent age at 20.0°C and activation energies determined in accordance with ASTM C1074. The curing method did not affect the temperature history of concrete significantly; therefore differences in compressive strength must be attributed to differences in the amount of moisture supplied to the specimens.

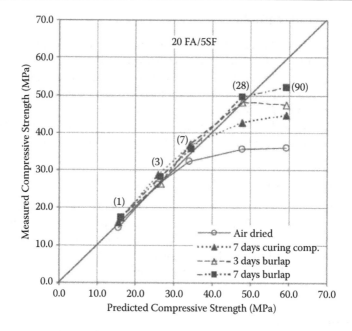

Figure 6.11 Effect of curing on strength development compared with strength predicted by maturity. (From Radlinski, M., Olek, J., and Nantung, T., *Transportation Research Record: Journal of the Transportation Research Board,* Transportation Research Board of the National Academies, Washington, DC, 2070, 49–58, 2008. With permission.)

Plots of predicted versus measured strengths show a strong dependency between curing and when the curve deviates from the line of equality (Figure 6.11).

This means that the maturity method can be successfully employed for strength prediction as long as the curing practices are satisfactory. In addition the data indicates that as the quality of curing is reduced, reliability of the method drops with increasing FA content.

Trial batch and bridge deck concrete

Two further tests were conducted in a trial batch and a mixture used in a bridge deck built in Indiana.

The mixtures contained 249 kg/m³ Type I cement; 66 kg/m³ of Class C FA; and 17 kg/m³ of SF. The water to cementitious materials ratio was 0.40.

A trial batch was used to develop the maturity function and to further assess the effect of curing. Three curing/drying conditions were applied: (a) continuous water curing; (b) air drying; and (c) air drying with intermittent moist curing comprising 6 hours in a moist room at 2, 5, and 9 days.

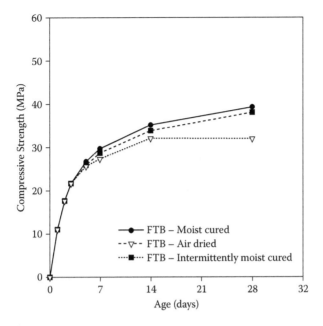

Figure 6.12 Effect of curing on compressive strength. (From Radlinski, M., Olek, J., and Nantung, T., *Transportation Research Record: Journal of the Transportation Research Board,* Transportation Research Board of the National Academies, Washington, DC, 2070, 49–58, 2008. With permission.)

The deck mixture specimens were left at the construction site for the first 24 hours, covered with wet burlap and plastic sheet in Styrofoam containers for 6 days, then exposed to the ambient environment. Temperatures were monitored in the cylinders and in the deck and were found to be similar. Compressive strengths were measured up to 90 days.

The compressive strength development of the trial concrete subjected to various curing conditions is presented in Figure 6.12. The concrete specimens subjected to air-drying ceased to develop strength after 14 days while the intermittent moisture samples were not significantly affected.

The accuracy of the prediction of both air-cured and intermittently moist-cured field concrete specimens was adequate up to 14 days.

The predicted compressive strength of the deck mixture is about 10 MPa lower than the actual strength. The observed strength discrepancies were the result of the differences in air content between mixture used to develop the model and the actual mixture used.

In summary:

- The ability of the maturity method to predict strength development of ternary concrete at ages greater than 7 days is significantly influenced by air content, initial curing condition and, to a lesser extent, by the composition of the mixture.

- Intermittent moist curing did not significantly reduce cylinder strengths.

Discussion

Two points warrant some discussion:

- The insensitivity of the temperature profile to curing is somewhat surprising. The authors tie this to the availability of water to the system, but water available for hydration should result in an increase in heat of hydration. On the other hand, a sample that is allowed to be tested dry would normally be expected to exhibit higher rather than lower strengths. It may be that the effect is present but smaller than the precision of the data collection.
- The effects of intermittent wetting should be considered with some care. The conclusions are based on measurement of compressive strength of 150 mm cylinders. As discussed in Chapter 3, strength is a relatively coarse indicator of curing effects, particularly in larger samples such as these. It is likely that measurement of permeability at the surfaces of the different samples would have led to a different conclusion.

The data once again clearly demonstrate the effects of curing practice on the long-term hydration of a concrete mixture.

Bouzoubaâ et al. (2010)

In other work Bouzoubaâ et al. (2010) looked at the effects of mixture properties on carbonation rates in accelerated laboratory and field tests.

Mixtures contained Type I portland cement and two Class F fly ashes at 0%, 20%, 35%, and 50%. Target strengths were 25, 35, and 45 MPa at 28 days. Samples were tested in compressive strength, accelerated carbonation, and natural carbonation indoor and outdoor. Moist (fog) curing was applied for 3, 7, and 10 days after which the samples were placed in the accelerated test. Accelerated carbonation was in a chamber with 3% CO_2 at 23°C and 65% relative humidity. Samples for the natural exposures were cured for 7 days before being exposed.

The depth of carbonation in the accelerated exposure was measured up to 140 days in accordance with the method described in RILEM Committee CPC-18 (1988). The depth of carbonation was measured in samples from the natural exposures for up to 4 years (Figure 6.13).

The data shows that the rate of carbonation increases with increasing fly ash content, decreasing strength, and to a far lesser extent with longer curing up to 7 days. Little benefit of longer curing was observed. It is notable

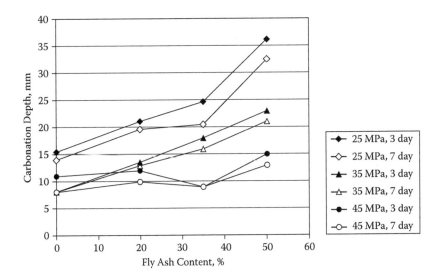

Figure 6.13 Effect of curing time and design strength on carbonation. (Data from Bouzoubaâ, N., Bilodeau, A., Tamtsia, B., and Foo, S., *Canadian Journal of Civil Engineering*, 37, 1535–1549, 2010.)

that as strength increases the effect of fly ash dosage decreases. This is likely due to the improved permeability of the system with lower w/cm.

The natural exposure data confirm that carbonation rates increase with increasing fly ash content and reducing strength. The rates were reportedly higher in the interior samples than the exterior, likely because the relative humidity was closer to that which promotes carbonation.

Discussion

Curing beyond 3 days appears to show limited benefit with respect to carbonation. It is also notable that carbonation was less sensitive to fly ash content in the higher-grade concrete mixtures.

Peyton et al. (2012)

Peyton et al. (2012) also investigated cracking of bridge decks. Five bridge decks under construction were examined in 2005 to determine which of these factors contributed to cracking:

- Structural design
- Material properties
- Mixture proportioning
- Construction practices
- Curing practices

The primary hardened properties investigated included permeability and compressive strength. The fresh concrete properties tested were slump, unit weight, air content, and concrete temperature.

Multiple samples were taken from various locations in the decks being evaluated.

- Compressive strengths were evaluated at 1, 7, 28, and 56 days.
- Shrinkage readings were taken at 1, 4, 7, 28, 56, and 112 days.
- Rapid chloride ion penetration (RCP) tests were conducted at 28 and 90 days.
- Freezing and thawing testing began at 14 days.

Deck cracking was assessed by visually locating and documenting cracks in terms of their length location, orientation, and approximate widths.

Concrete used in bridge decks in Arkansas is required to have

- Minimum 28-day compressive strength of 28 MPa
- Slump of 25 to 100 mm
- Air content of 4% to 8%
- Maximum w/cm of 0.44
- Minimum cementitious material content of 362 kg/m³
- Maximum fly ash replacement rate of 20%
- Maximum slag replacement rate of 25% by weight

The specifications allowed the following curing approaches:

- Burlap–polyethylene sheeting
- Polyethylene sheeting or copolymer/synthetic blanket
- Membrane curing compounds

The bridge deck must be covered immediately after finishing and remain covered for at least 7 days. The curing materials must be kept wet (except for membrane curing).

All five bridge decks were continuously poured in the late evening or early morning. In all cases the concrete was pumped up to the deck, fogged during placement, screeded, floated, tined, and then cured with curing compound and an absorbent cloth and plastic sheeting.

Cracking was mapped on all of the decks between 5 and 12 months after placement.

Bridge Deck 1—The cracks ranged from 100 mm to 14.6 m in length and from less than 0.127 mm to 0.61 mm in width. The cracks were a network of transverse and longitudinal cracks with diagonal cracks connecting them. Cracks were concentrated over the center support, possibly due to traffic vibrations and low early compressive strengths.

Bridge Deck 2—The cracks ranged from 76 mm to 5 m in length and 0.05 mm to 0.40 mm in width. The cracks were mostly transverse and heavily concentrated in the positive moment section near the piers. Plastic shrinkage cracks were noted where paste had been transported to during screeding.

Bridge Deck 3—The cracks ranged from 0.9 to 3.7 m in length and less than 0.18 mm wide. The cracks were mostly transverse cracks that started and stopped near beam lines.

Bridge Deck 4—There was very little cracking in the deck. The cracks ranged from 152 mm to 1.2 m in length and 0.05 to 0.25 mm in width.

Bridge Deck 5—The cracks were 0.60 to 1.50 m in length and 0.05 to 0.18 mm wide. Some were at 45 degree angles to the intermediate bents.

At the time of construction the following was noted:

- Four of the five bridge decks had slumps that exceeded specifications in at least one location.
- Three of the five bridge decks had measured air contents that did not meet specifications in at least one location.
- Two bridge decks had fresh concrete temperatures that were greater than allowed in at least one location.

Fresh properties were plotted versus the crack density for each deck and little correlation was observed (Figure 6.14). Hardened properties were

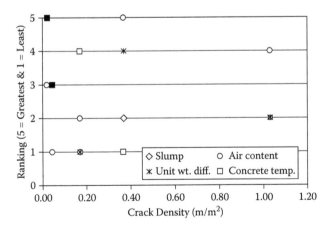

Figure 6.14 Ranking of fresh properties of the various bridge decks with respect to crack density. (From Peyton, S. W., Sanders, C. L., John, E. E., and Hale, W. M., *Construction and Building Materials*, 34, 70–76, 2012. With permission.)

plotted versus the crack density for each deck and little correlation was observed (Figure 6.15).

Although all curing activities complied with the specifications, there were differences in when activities were started. These differences were plotted against crack density. Some correlation is apparent that later application of curing was related to cracking density (Figure 6.16).

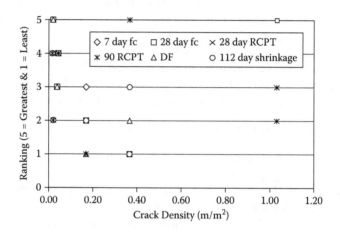

Figure 6.15 Ranking of hardened properties of the various bridge decks with respect to crack density. (From Peyton, S. W., Sanders, C. L., John, E. E., and Hale, W. M., *Construction and Building Materials*, 34, 70–76, 2012. With permission.)

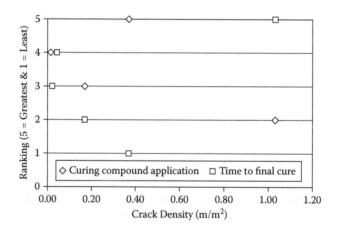

Figure 6.16 Ranking of curing method of the various bridge decks with respect to crack density. (From Peyton, S. W., Sanders, C. L., John, E. E., and Hale, W. M., *Construction and Building Materials*, 34, 70–76, 2012. With permission.)

Discussion

It is notable that all the bridges were cured with both curing compound and wet burlap and plastic, yet some cracking was still observed, especially those in which the curing was initiated late. This emphasizes the need to time curing correctly to obtain the most benefit.

Tamayo (2012)

The objective of work reported by Tamayo (2012) was to assess whether lithium silicate compounds affected the risk of plastic shrinkage crack-ing in four bridge decks in Arkansas. Part of each deck was treated with lithium silicate used as a finishing aid, while the whole deck was treated with conventional curing compound. Seal coats were later applied: lin-seed on the control section and lithium silicate for the test section. Data collected at time of construction included temperature, relative humidity, wind speed, and the evaporation rate. The decks were inspected to evalu-ate plastic shrinkage cracking.

Placement details are shown in Table 6.6.

Distress surveys were conducted on each deck on completion, after one year and another some time later. The total crack lengths recorded in the surveys is shown in Table 6.7.

The lithium silicate appeared to perform better than the conventional approach in all of the bridges assessed. In addition, labor requirements for application were lower, leading to similar final costs.

Discussion

The lithium silicate was used as for the purpose of reducing cracking, and not as a curing agent, but as such appears to have been effective and cost effective.

Table 6.6 Weather Conditions at Time of Placing Decks

Bridge	Month Placed	Air Temperature (°F)	Wind Speed (mph)	RH (%)	Evaporation (lb/ft²/hr)
1	September	63	0.8	78	0.020
2	September	74	4.5	52	0.085
3	March	80	20	33	0.300
4	August	89	1.4	65	0.054

Source: Data from Tamayo, S., "Evaluation of High Performance Curing Compounds on Freshly Poured Bridge Decks," TRC 1002, Arkansas State Highway and Transportation Department, 2012.

Table 6.7 Extent of Cracking in Bridge Decks[a]

Survey Time	Lithium Silicate	Conventional
Bridge 1		
New	277	300[b]
1 year	672	1190
Final	1750	2654
Bridge 2		
New	173	301
1 year	—	—
Final	1402	1608
Bridge 3		
New	236	—
1 year	1208	—
Final	2485	—
Bridge 4		
New	265	327
1 year	906	1137
Final	1709	1978

Source: Data from Tamayo, S., "Evaluation of High Performance Curing Compounds on Freshly Poured Bridge Decks," Arkansas State Highway and Transportation Department, TRC 1002, 2012.

[a] In linear feet.
[b] Incomplete measurement because equipment covered a portion of the deck.

SPECIFICATIONS AND TESTING

Choi et al. (2012)

Choi et al. (2012) conducted fieldwork to assess real-time application rates of curing compound on pavements. The work conducted sought to identify methods to measure application rates and to implement procedures in practice.

The test methods initially evaluated included tests on the concrete's

- Penetration resistance
- Initial surface adsorption
- Surface temperature
- Reflectance
- Relative humidity
- Dielectric constant

It was concluded that none of the methods were practical nor accurate enough to be useful for acceptance testing.

Attention was then turned to measuring curing cart speed and spray rates, which are the dominant factors controlling application rates. Spray rates are a direct function of pressures in the system for a given set of nozzles. Pressures are not normally adjusted; therefore the single control parameter is cart speed.

The speed of a curing cart is normally controlled by the operator who is seeking either to deliver a uniform color based on what they can see, or just to catch up with the paving machine without running off the road. Reportedly operators are not often provided guidance on what is needed from what they are doing.

The research team sought to find and implement a practical speed sensor system and selected a noncontact Doppler radar speed sensor (NDRSS) connected wirelessly to a data logger that could be accessed by contractor and agency for both quality control and acceptance purposes (Figure 6.17).

The flow rate of the curing compound was estimated directly by weighing the amount of curing compounds collected in a bucket from one nozzle over a fixed time.

Verification of the testing procedure was conducted weighing plates before and after they were sprayed by the curing cart. The data indicated that the approach of measuring curing cart speed could be used with sufficient accuracy for acceptance purposes. Differences between measured and calculated rates were attributed to losses due to wind.

An attempt was made to quantify this effect based on the wind speed and there was a trend, but the scatter was large. The data indicated that

Figure 6.17 Doppler radar speed sensor mounted on a curing cart. (From Choi, S., Yeon, J. H., and Won, M. C., *Construction and Building Materials*, 35, 597–604, 2012. With permission.)

cart speed had to be reduced by roughly 0.65% for each 1 km/h of wind to maintain required application rates.

The procedure recommended was that the speed control on any cart should be calibrated periodically for the nozzles fitted and the required coverage using the Doppler device. This may mean a red line being marked on the control knob, which should not be passed in operation.

Discussion

The procedure recommended appears to find a reasonable balance between practicality and accuracy, although the compensation for wind, other than the use of windshields, may need more attention.

Kropp et al. (2012)

The Wisconsin Department of Transportation (WisDOT) conducted a project that set out performance characteristics for curing compounds that will be used in cold weather. Parameters investigated include salt scaling, chloride ion penetration, moisture loss, and carbonation.

Mixtures tested contained portland cement with 30% slag or 30% Class C fly ash. The w/cm was 0.40 for the OPC and slag mixtures and 0.37 for the fly ash mixtures to maintain constant slump near 3 inches. Air content was controlled to about 6%. Two aggregate types were included: gravel and limestone. Each mixture was treated with a range of curing compounds applied to the samples using handheld sprayers 2 hours after mixing. The application rate of 200 ft²/gallon for all compounds was monitored by weighing the samples before and after spraying.

After spraying, specimens used for the salt scaling tests were exposed to the atmosphere for 28 days at 45% RH and 23°C followed by 14 days at 50% and 23°C. Testing was conducted in accordance with ASTM C672.

Chloride penetration was evaluated using the ponding method in AASHTO T259.

Accelerated carbonation was assessed on mortar specimens sprayed with curing compound, then stored for 3 days at 100°F and 32% RH and finally placed in a 100% CO_2 atmosphere for 1 month. Samples were then split and visually assessed using phenolphthalein solution.

Some samples used for the AASHTO T259 testing were also sawed into 1 cm cubes and exposed to deicer solution for a period of 3 months at constant temperature then examined in a scanning electron microscope (SEM).

Curing products used included the following:

- Wax-based emulsion (cure)
- Linseed oil emulsion (cure)
- Poly(alpha-methylstyrene) (PAMS) resin (cure)

- Chlorinated rubber epoxy (cure and seal)
- Clear acrylic

Analysis of the salt scaling data showed no clear trend between curing approaches across all the binder/aggregate combinations, except that the linseed was the worst performer in every case (Figure 6.18). This was attributed to differences in surface conditions at the time the curing compounds were applied.

For the chloride penetration tests there was little variation between cement or curing compound types.

Ranking of the rate of evaporative moisture loss as a function of curing type was (from fastest to slowest):

- None
- Wax
- Linseed
- PAMS
- Acrylic
- Chlorinated rubber

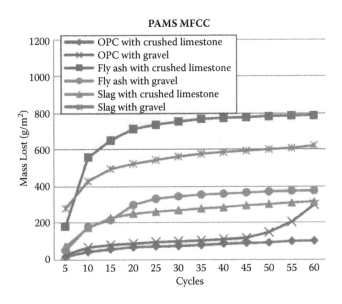

Figure 6.18 Typical plots of mass loss under salt scaling testing for various mixtures using PAMS curing compound. (From Kropp, R., Cramer, S. M., and Anderson, M. A., "Laboratory Study of High Performance Curing Compounds for Concrete Pavement," Wisconsin Department of Transportation, SPR #0092-11-05, 2012. With permission.)

Ranking of the rate of carbonation as a function of curing type was as follows (from fastest to slowest):

- Linseed
- PAMS
- Wax
- None
- Acrylic
- Chlorinated rubber

Overall findings include the following:

- Curing compounds were less effective than cure-and-seal products in the tests conducted.
- Linseed oil did not perform as well as the other products.

It was noted that the performance of the products is likely strongly related to the moisture state of the surface when the product is applied emphasizing the importance of confirming that bleeding has stopped before spraying.

A weighted decision matrix was developed to assist designers in specifying curing compound type (Table 6.8).

Discussion

The poor performance of the PAMS product is surprising and may be related to the discussion in bleeding. Similarly the variation in performance of the curing products with changing SCM type is likely because SCMs affect bleeding rate and duration (Taylor et al. 2004).

SUMMARY

In summary, the following seem to be common themes that emerge from the works discussed:

- Curing is critically important for structures exposed to severe environments, particularly from the point of view of potential durability.
- In arid areas it may be advisable to avoid self-desiccating mixtures because added water approaches may be difficult to achieve.
- Delayed application of curing measures can markedly reduce potential durability.
- Some papers reported better performance from curing compounds, while others pointed to the superiority of burlap and plastic.

Table 6.8 Guide to Selecting Curing Type (Low Value = High Performance)

Aggregate Type	Cement Type	Property	Coating Type				
			PAMS	Linseed	Acrylic	Chlorinated Rubber	Wax
Limestone	OPC	Scaling	3	3	1	1	3
		Chloride Penetration Resistance	2	2	2	3	3
	30% Fly Ash	Scaling	4	5	3	5	2
		Chloride Penetration Resistance	2	2	1	2	3
	30% Slag	Scaling	2	10	2	2	5
		Chloride Penetration Resistance	1	2	1	1	3
Gravel	OPC	Scaling	2	10	2	2	1
		Chloride Penetration Resistance	1	1	1	1	2
	30% Fly Ash	Scaling	2	5	2	4	1
		Chloride Penetration Resistance	1	1	1	1	1
	30% Slag	Scaling	4	5	2	2	5
		Chloride Penetration Resistance	1	2	2	1	2
		Total	**14**	**29**	**10**	**14**	**16**
		Weighted Average	**2.1**	**4.0**	**1.6**	**2.1**	**2.6**

Source: From Kropp, R., Cramer, S. M., and Anderson, M. A., "Laboratory Study of High Performance Curing Compounds for Concrete Pavement," Wisconsin Department of Transportation, SPR # 0092-11-05, 2012. With permission.

- Application of curing compounds before bleeding has ended is likely to compromise performance.
- Curing compounds appear to provide protection equivalent to between 3 and 7 days moist curing.
- Curing is required for 3 to 7 days for most performance parameters, with the longer periods needed for systems containing SCMs.
- Steam curing may accelerate strength gain but compromise potential durability.
- Effective curing can reduce the risk of cracking.
- Lithium silicate products appear to be effective at reducing cracking in bridge decks.
- Calibrating and controlling curing cart speed appear to have a beneficial effect on curing effectiveness.

REFERENCES

Al-Gahtani, A. S., 2010, "Effect of Curing Methods on the Properties of Plain and Blended Cement Concretes," *Construction and Building Materials*, vol. 24, pp. 308–314.

Alsayed, S. H., and Amjad, M. A., 1994, "Effect of Curing Conditions on Strength, Porosity, Absorptivity, and Shrinkage of Concrete in Hot and Dry Climate," *Cement and Concrete Research*, vol. 24, no. 7, pp. 1390–1398.

Bouzoubaâ, N., Bilodeau, A., Fournier, B., Hooton, R. D., Gagné, R., and Jolin, M., 2008, "Deicing Salt Scaling Resistance of Concrete Incorporating Supplementary Cementing Materials: Laboratory and Field Test Data," *Canadian Journal of Civil Engineering*, vol. 35, no. 11, pp. 1261–1275.

Bouzoubaâ, N., Bilodeau, A., Fournier, B., Hooton, R. D., Gagné, R., and Jolin, M., 2011, "Deicing Salt Scaling Resistance of Concrete Incorporating Fly Ash and (or) Silica Fume: Laboratory and Field Sidewalk Test Data," *Canadian Journal of Civil Engineering*, vol. 38, pp. 373–382.

Bouzoubaâ, N., Bilodeau, A., Tamtsia, B., and Foo, S., 2010, Carbonation of Fly Ash Concrete: Laboratory and Field Data," *Canadian Journal of Civil Engineering*, vol. 37, pp. 1535–1549.

Boyd, A. J., and Hooton, R. D., 2007, "Long-Term Scaling Performance of Concretes Containing Supplementary Cementing Materials," *Journal of Materials in Civil Engineering*, vol. 19, pp. 820–825.

Choi, S., Yeon, J. H., and Won, M. C., 2012, "Improvements of Curing Operations for Portland Cement Concrete Pavement," *Construction and Building Materials*, vol. 35, pp. 597–604.

Hoppe, G. E., Mackechnie, J. R., and Alexander, M., 1994, "Measures to Ensure Concrete Durability and Effective Curing During Construction," Report No. RR 93/463, South African Department of Transport, Pretoria.

Huo, X. S., and Wong, L. U., 2006, "Experimental Study of Early-Age Behavior of High Performance Concrete Deck Slabs under Different Curing Methods," *Construction and Building Materials*, vol. 20, pp. 1049–1056.

Johnston, D. P., and Surdahl, R. W., 2007, "Influence of Mixture Design and Environmental Factors on Continuously Reinforced Concrete Pavement Cracking," *Transportation Research Record: Journal of the Transportation Research Board,* No. 2020, pp. 83–88.

Kropp, R., Cramer, S. M., and Anderson, M. A., 2012, "Laboratory Study of High Performance Curing Compounds for Concrete Pavement," SPR #0092-11-05, Wisconsin Department of Transportation.

Petrou, M. F., Harries, K. A., and Schroeder, G. E., 2001, "Field Investigation of High-Performance Concrete Bridge Decks in South Carolina," Transportation Research Record 1770, Paper No. 01-0420.

Peyton, S. W., Sanders, C. L., John, E. E., and Hale, W. M., 2012, "Bridge Deck Cracking: A Field Study on Concrete Placement, Curing, and Performance," *Construction and Building Materials*, vol. 34, pp. 70–76.

Poursaee, A., and Hansson, C. M., 2010, "Curing Time and Behaviour of High-Performance Concrete," *Proceedings of the Institution of Civil Engineers Construction Materials*, vol. 163, no. 4, pp. 223–230.

Radlinski, M., Olek, J., and Nantung, T., 2008, "Influence of Curing Conditions on Strength Development and Strength Predictive Capability of Maturity Method," *Transportation Research Record: Journal of the Transportation Research Board,* no. 2070, pp. 49–58.

RILEM, 1988, "CPC-18: Measurement of Hardened Concrete Carbonation Depth," *Materials and Structures,* vol. 21, no. 126, pp. 453–455.

Tamayo, S., 2012, "Evaluation of High Performance Curing Compounds on Freshly Poured Bridge Decks," TRC 1002, Arkansas State Highway and Transportation Department.

Taylor, P. C. Morrison, W., and. Jennings, V. A., 2004, "The Effect of Finishing Practices on Performance of Concrete Containing Slag and Fly Ash as Measured by ASTM C 672 Resistance to Deicer Scaling Tests," *Cement, Concrete, and Aggregates,* vol. 26, no. 2., pp. 155–159.

Zhimin, H. E., and Junzhe, L. I. U., 2011, "Effect of Steam Curing on Water Sorptivity of Concrete," *Key Engineering Materials*, vol. 477, pp. 263–267.

Index